Lecture Notes in Mathematics

Edited by A. Dold and B. Eckmann

1031

Dynamics and Processes

Proceedings of the Third Encounter in Mathematics and Physics, held in Bielefeld, Germany
Nov. 30 – Dec. 4, 1981

Edited by Ph. Blanchard and L. Streit

Springer-Verlag
Berlin Heidelberg New York Tokyo 1983

Editors

Ph. Blanchard
L. Streit
Theoretische Physik
Fakultät für Physik, Universität Bielefeld
4800 Bielefeld 1, Federal Republic of Germany

AMS Subject Classifications (1980): 46L, 58F, 60B, 70F, 76, 81C, 81E, 83C

ISBN 3-540-12705-4 Springer-Verlag Berlin Heidelberg New York Tokyo
ISBN 0-387-12705-4 Springer-Verlag New York Heidelberg Berlin Tokyo

This work is subject to copyright. All rights are reserved, whether the whole or part of the material is concerned, specifically those of translation, reprinting, re-use of illustrations, broadcasting, reproduction by photocopying machine or similar means, and storage in data banks. Under § 54 of the German Copyright Law where copies are made for other than private use, a fee is payable to "Verwertungsgesellschaft Wort", Munich.

© by Springer-Verlag Berlin Heidelberg 1983
Printed in Germany

Printing and binding: Beltz Offsetdruck, Hemsbach/Bergstr.
2146/3140-543210

Mathematicians as well as physicists have had to travel great distances and overcome amazing obstacles in order to gather once again for fruitful collaboration.

Res Jost, Varenna 1968

We dedicate this volume to the wisdom and perseverance of our travel guide.

Ph. Blanchard, L. Streit

INTRODUCTION

The present volume collects the Proceedings from the IIIrd session of the Bielefeld Encounters in Physics and Mathematics: "Quelques points de contact entre Mathématiques et Physique en Allemagne Fédérale et en France" that took place at the Center for Interdisciplinary Research (ZiF) of Bielefeld University November 30 to December 4,1981.

It was in the tradition of the other Bielefeld Encounters in 1976,1978 and 1982 and as well in the tradition of a close French-German collaboration under the auspices of a treaty between the UER of Marseilles-Luminy and the Physics Department of Bielefeld University. This collaboration began when a large group of French mathematicians and physicists were in residence at the Center for Interdisciplinary Research in the academic year 1975-1976. At that time the CIRM (Centre International de Rencontres Mathématiques) of the French Mathematical Society was in an early planning stage, but already then there was the suggestion that CIRM and ZiF might serve as a "pair of homes" for French-German scientific exchange on the interdisciplinary interface between mathematics and physics. Very recently CIRM has indeed hosted the second meeting of this kind, and both sides hope that this marks the beginning of a tradition of exchange and collaboration.

It was of great help to have the support of the French Embassy in Bonn, the DAAD, the Westfälisch-Lippische Universitätsgesellschaft and of Ministerpräsident Dr. B. Vogel as Federal Coordinator of cultural relation in the framework of the treaty for French-German cultural cooperation.

List of participants

S.Albeverio
T.Arede
A.Aragnol
K.Baumann
Ph.Blanchard
R.Böhme
H.J.Borchers
E.Brüning
Ph.Combe
J.Cuntz
S.Doplicher
A.Dress
J.Ehlers
V.Enß
D.Kastler
W.Krüger
J.Lascoux
J.Leray
C.Macedo
M.Mebkhout
D.Miller
J.Potthoff
J.Quaegebeur
J.Rezende
J.E.Roberts
R.Rodriguez
R.Sénéor
H.Rost
M.Sirugue-Collin
M.Sirugue
J.M.Souriau
R.Schrader
G.Sommer
O.Steinmann
L.Streit
D.Testard
J.L.Verdier
T.Yoshimura
E.Zehnder

C O N T E N T S

Symmetries and Covariant Representations

H.-J. Borchers
Institut für Theoretische Physik der
Universität Göttingen

I. Introduction and notations

Symmetries are one of the most powerful concepts in physics. Many of the classification schemes of physical objects are based on symmetry-groups. Therefore it is no wonder that one finds a vast amount of literature on this subject. In earlier times most of these investigations have been focused on the classification of group representations. Since the last 1 1/2 decade, however, the interest has changed more to the investigation of the interplay between the symmetry-group and the algebra of observables or the field-algebra. This subject runs now under the name of C*-dynamical systems. So far the main tool of this subject is, what has been called by Doplicher, Kastler, and Robinson (9) the covariance-algebra, and which is now known as crossed product between a C*-algebra and a group. For a good survey on this subject see the book of G.K. Petersen (11) where one finds also a list of references. One of the objects of this theory is to characterize the representations in which we have also a continuous representation of the symmetry-group implementing the automorphisms. This problem has been answered modulo problems of multiplicity by the author (3) (4).

Looking at this part of the theory of C*-dynamical systems one finds that there are two assumptions which are unsatisfactory. The first assumption is the continuity assumption which says that the expressions $g \rightarrow \alpha_g(x)$ have to be continuous functions on the group with values in the C*-algebra furnished with the norm topology. However, in quantum field theory or statistical mechanics one usually starts with a local net of von Neumann algebras, which means that the continuity assumption is violated. The standard argument out of this dilemma is usually that one postulates the existence of a sub-algebra fulfilling the continuity assumption. Since such a sub-algebra, if it exists, will mostly not be dense in the norm topology one cannot be sure wheter one investigates the original object, in particular when different representations are involved. Looking at the existing theory one knows that not the norm topology on the algebra A is important for the representation of the group, but the norm topology on the dual-space A* of A (see e.g.(3)). The assumption of the continuity of the action of G on A is therefore only a technical assumption used in order to have an easy access to the problem of covariant representation.

The existing theory has a second defect, namely it can handle only locally compact groups and not arbitrary topological groups. The restriction to the case of locally compact groups was again dictated by mathematical convenience because the mathematical theory for such groups is existing, but not for general topological groups. At that time where covariance algebras have been invented, it was enough to study locally compact groups, because all global symmetry-groups in physics are of this nature. During the last years the theory of gauge-fields is developing with high speed and they hopefully will enter into axiomatic very soon. If this is the case then we will have to deal with topological groups which are no longer locally compact.

Because of these arguments, I feel it is necessary to study the problem of covariant representations again in a more general setting. This will be done in this and a forthcoming paper. We will assume that we are dealing with a C*-dynamical system (A,G,α), this is a C*-algebra A, a topological group G and a mapping α of G into Aut(A). It is no longer assumed that the action α_g on A is continuous in any topology on A. As mentioned before what counts is the continuity property of the group action on the dual-space A* of A. The action of G on A* should be denoted by α'_g or α^*_g , but since there is usually no confusion possible we will denote the transposed and the double transposed of α_g again by α_g. We will denote by A^*_c the set of functionals $\phi \in A^*$ such that the function $g \to \alpha_g\phi$ is a continuous function on G with values in A* furnished with the norm topology.

Since the dual-space A* of a C*-algebra is at the same time the predual of the eveloping von Neumann algebra A** and since our object of investigation is a subspace of A* it turns out that we have exclusively to deal with the dual pair (A*,A**). Therefore it is natural to forget about the original C*-algebra A and work in the theory of von Neumann algebras. This means we will deal with (M_*,M) where M is a von Neumann algebra and M_* its pre-dual.

Let $M_{*,c}$ denote that part of M_* on which the group acts strongly continuous. $M_{*,c}$ is a norm-closed linear subspace of M_*. Given two positive linear functionals $\omega_1,\omega_2 \in M_{*,c}$ we construct an interpolating functional ϕ_{ω_1,ω_2} belonging again to $M_{*,c}$ and which fulfills some further useful requirements. This is done in the next section with the help of the so-called natural cone which is constructed using Tomita's theory of modular Hilbert algebra. This interpolating functional is used in the third section for constructing covariant representations and proving the main results. Some remarks on open problems will end these notes.

II. On interpolating functionals

For two given normal positive linear functionals $\omega_1,\omega_2 \in M_*^+$ one often needs another linear functional $\phi \in M_*$ with the property $|\phi(x^*y)| \leq \omega_1(x^*x)\omega_2(y^*y)$. There exists of course many such functionals but usually one wants some additional nice properties of such functionals.

In particular one would like to establish a map from $M_*^+ \times M_*^+ \to M_*$ which is continuous in the norm topology and which commutes with all normal automorphisms of M. That such a map exists is the content of this section. We want to show:

II.1. Theorem

Let M be a von Neumann algebra with pre-dual M_*. Then to every pair $\omega_1,\omega_2 \in M_*^+$ exists a linear functional $\phi_{\omega_1,\omega_2} \in M_*$ with the properties:

a) $\phi_{\omega,\omega} = \omega$

b) For $\omega_i \in M_*^+$ and $x_i \in M$, $i = 1,\dots,n$

we have $\sum_{i,j} \phi_{\omega_i,\omega_j}(x_i^* x_j) \geq 0$.

c) For every $\alpha \in \mathrm{Aut}(M)$ we have $\alpha\,\phi_{\omega_1,\omega_2} = \phi_{\alpha\omega_1,\alpha\omega_2}$.

d) For $\omega_1,\dots\omega_4 \in M_*^+$ we get the estimate

$$||\phi_{\omega_1,\omega_2} - \phi_{\omega_3,\omega_4}|| \leq ||\omega_1 - \omega_3||^{\frac{1}{2}}||\omega_2||^{\frac{1}{2}} + ||\omega_2 - \omega_4||^{\frac{1}{2}}||\omega_3||^{\frac{1}{2}}$$

and

$$||\phi_{\omega_1,\omega_2} - \phi_{\omega_3,\omega_4}|| \leq ||\omega_1 - \omega_3||^{\frac{1}{2}}||\omega_4||^{\frac{1}{2}} + ||\omega_2 - \omega_4||^{\frac{1}{2}}||\omega_1||^{\frac{1}{2}}.$$

e) $\phi^{*}_{\omega_1,\omega_2} = \phi_{\omega_2,\omega_1}$

This theorem will be proved with the help of Tomita's theory of modular Hilbert algebras (14) (see e.g. M.Takesaki (19)) for a representation of this subject. In particular the theory of the so-called natural cone and standard representation of positive functionals is needed. This theory of the natural cone has been developed by Araki (1),(2), Connes (7),(8), and Haagerup (10). In the papers of Araki and Connes one finds the case where the von Neumann algebra M has a seperating state. The general case which uses weights instead of states is treated by Haagerup. Since we have in mind that our von Neumann algebra is the double dual of a C*-algebra we need the general case, since except for special situations the double dual of a C*-algebra will not have seperating states. For an introduction into the theory of the natural self-dual cone see e.g. the textbook of Bratteli and Robinson (6) Vol. I., section 2.5.4. for the case where M has a separating normal state, and section 2.7.3. for the general situation.

Let w be a faithful, normal, semi-finite weight on M and L_w be the left ideal such that $w(x^*x) < \infty$, the w defines a scalar product on L_w. Let H_w be the completion of L_w in this scalar product and $\eta : L_w \to H_w$ the canonical embedding of L_w. Since L_w is a left ideal one can construct a representation π_w of M on H_w which is faithful and normal, since w was a faithful normal weight. In the same way as in the case where one has a separating state it is possible to construct the modular operator Δ and the modular involution J. If one defines the algebra $\mathfrak{M}_w = L_w^* L_w$ then this is ultraweakly dense in M. Now the natural cone P is defined by

$$\Delta^{\frac{1}{4}}\, \eta(\mathfrak{M}_w^+) = \overline{\{\pi_w(x) J \eta(x) ; x \in \mathfrak{M}_w\}}$$

where the bar denotes the closure in the Hilbert space H_w.

The importance of the cone P is the following: For every positive linear functional $\omega \in M_*^+$ exists a unique vector $\xi_\omega \in P$, such that $\omega(x) = (\xi_\omega, \pi_w(x)\xi_\omega)$ holds. In addition the map $\omega \to \xi_\omega$ is bi-continuous. More precisely one finds the estimate

$$||\xi_{\omega_1} - \xi_{\omega_2}||^2 \leq ||\omega_1 - \omega_2|| \leq ||\xi_{\omega_1} - \xi_{\omega_2}||\ ||\xi_{\omega_1} + \xi_{\omega_2}|| \quad (*)$$

Knowing that every positive linear functional has a unique representative in natural cone P we can define interpolating functionals.

II.2. Definition

Let M be a von Neumann algebra, w a faithful mormal semi-finite weight on M, $\{\pi_w, H_w\}$ its canonical representation and P the natural cone defined by w. For $\omega_1, \omega_2 \in M_*^+$ we put

$$\phi_{\omega_1,\omega_2}(x) = (\xi_{\omega_1}, \pi_w(x)\xi_{\omega_2})$$

where ξ_{ω_1} and ξ_{ω_2} are the unique representations of ω_1 and ω_2

II.3. Lemma

Definition II.2. together with the above estimate (*) imply the conditions a, b and e of theorem II.1.

Proof: a follows directly from the definition II.2., and so does e.

b) Let $\omega_i \in M_*^+$ and $x_i \in M$, $\quad i = 1, \ldots, n$
then we get:

$$\sum_{i,j} \phi_{\omega_i,\omega_j}(x_i^* x_j) = \sum_{i,j} (\xi_{\omega_i}, \pi_w(x_i^* x_j)\xi_{\omega_j}) = ||\sum_j \pi_w(x_j)\xi_{\omega_j}||^2 \geq 0$$

d) We have:

$$|\phi_{\omega_1,\omega_2}(x) - \phi_{\omega_3,\omega_4}(x)| = |(\xi_{\omega_1}, \pi_w(x)\xi_{\omega_2}) - (\xi_{\omega_3}, \pi_w(x)\xi_{\omega_4})|$$

$$= |(\xi_{\omega_1} - \xi_{\omega_3}, \pi_w(x)\xi_{\omega_2}) - (\xi_{\omega_3}, \pi_w(x)(\xi_{\omega_3} - \xi_{\omega_4}))|$$

$$\varepsilon\ ||x||\{||\xi_{\omega_1} - \xi_{\omega_3}||\ ||\xi_{\omega_2}|| + ||\xi_3||\ ||\xi_{\omega_2} - \xi_{\omega_4}||\}$$

$$\leq ||x||\{||\omega_1 - \omega_3||^{\frac{1}{2}}||\omega_2||^{\frac{1}{2}} + ||\omega_3||^{\frac{1}{2}}||\omega_2 - \omega_4||^{\frac{1}{2}}\}$$

This gives the first inequality. The second inequality follows in the same way or by symmetry.

Before proving the last statement of the theorem II.1. we have to mention some further known properties of the natural cone. These properties are:

(α) Every element in P is invariant under the involution J.

(β) P generates the Hilbert-space H. More precisely, if $\eta \in H_W$ and $J\eta = \eta$ then exist $\xi_1, \xi_2 \in P$ with $\xi_1 \quad \xi_2$ and $\eta = \xi_1 - \xi_2$

(γ) P is self dual, this means, if $\eta \in H_W$ and $(\eta, \xi) \geq 0$ for all $\xi \in P$ then follows $\eta \in P$.

(δ) These properties stay true for any closed face of P and the sub-Hilbert-space generated by this face.

(ε) The most remarkable feature of the natural cone P is the following. Let $\alpha \in \mathrm{Aut}(M)$ define a map $V: P \to P$ by the relation $V\xi_\omega = \xi_{\alpha\omega}$ then it turns out that this map is linear and that it can be extended to a unitary operator on all of H_W. Moreover one has the relation

$$V\pi_W(x)V^* = \pi_W(\alpha^{-1}(x)).$$

<u>Proof of the theorem:</u> In Lemma II.3. we have proved all parts of the theorem except statement c). This is now a consequence of the previous remarks on the natural cone P by the following equation

$$\phi_{\alpha\omega_1,\alpha\omega_2}(x) = (\xi_{\alpha\omega_1}, \pi_W(x)\xi_{\alpha\omega_2}) = (V\xi_{\omega_1}, \pi_W(x)V\xi_{\omega_2})$$

$$= (\xi_{\omega_1}, V^*\pi_W(x)V\xi_{\omega_2}) = (\xi_{\omega_1}, \pi_W(\alpha x)\xi_{\omega_2}) = (\alpha\phi_{\omega_1,\omega_2})(x).$$

Next we want to draw some conclusions for our problem.

<u>II.4. Proposition</u>

Let $\{M,G,\alpha\}$ be a W*-dynamical system with G a topological group. Then we find:

a) For $\omega_1, \omega_2 \in M^+_{*,c}$ follow $\phi_{\omega_1,\omega_2} \in M_{*,c}$

b) Let w be a faithful, normal, semi-finite weight on M, and let P be the natural cone defined by w, then the set of representatives $\{\xi_\omega, \omega \in M^+_{*,c}\}$ which we denote by P_c is a closed sub-cone of P.

Proof:

a) If ω_1 or ω_2 is zero then also $\phi_{\omega_1,\omega_2} = 0$. Hence we can assume $\omega_1 \neq 0$ and $\omega_2 \neq 0$. Since $\omega_1, \omega_2 \in M^+_{*,c}$ exist for every $\varepsilon > 0$ a neighbourhood U_ε of the identity in G with

$$||\omega_1 - \alpha_g\omega_2|| < \frac{\varepsilon^2}{4||\omega_2||} \quad \text{and} \quad ||\omega_2 - \alpha_g\omega_2|| < \frac{\varepsilon^2}{4||\omega_1||} \quad \text{for } g \in U_\varepsilon.$$

Hence we obtain by theorem II.1. for $g \in U_\varepsilon$

$$||\phi_{\omega_1,\omega_2} - \alpha_g\phi_{\omega_1,\omega_2}|| = ||\phi_{\omega_1,\omega_2} - \phi_{\alpha_g\omega_1,\alpha_g\omega_2}||$$

$$\leq ||\omega_1 - \alpha_g\omega_1||^{\frac{1}{2}}||\omega_2||^{\frac{1}{2}} + ||\omega_2 - \alpha_g\omega_2||^{\frac{1}{2}}||\omega_1||^{\frac{1}{2}} < \varepsilon$$

This shows $\phi_{\omega_1,\omega_2} \in M_{*,c}$

b) Let $\omega_1, \omega_2 \in M^+_{*,c}$ and $\xi_{\omega_1}, \xi_{\omega_2}$ be their representatives in the natural cone P. Then we obtain by the definition of the interpolating functionals ϕ_{ω_1,ω_2} the equation

$$\omega_{\xi_{\omega_1} + \xi_{\omega_2}} = \omega_1 + \omega_2 + \phi_{\omega_1,\omega_2} + \phi_{\omega_2,\omega_1} \in M^+_{*,c}$$

by part a). This shows P_c is a cone because

$$\omega_{\lambda\xi_\omega} = \lambda^2\omega \text{ for } \lambda \geq 0$$

Since the map $M^+_* \to P$ is a bijective homeomorphism follows from the fact that $M^+_{*,c}$ is closed in norm and that P_c is a closed set and hence a closed cone.

III. Quasi-covariant representations

In this and the following section we will work with C*-algebras. Everything which is said here stays true in the context of von Neumann algebras if one replaces the concept of representation by that of normal representation.

Let A be a C*-algebra, then we denote by S(A) the set of states. If (π,H) is a representation of A (which always will assumed to be non-degenerate without further mentioning) then we denote by F_π the folium of A. These are the normal states of $\pi(A)$. F_π is convex and norm closed and invariant under the map $\omega \to x\omega x^*/\omega(xx^*)$ for all $x \in A$ with $\omega(xx^*) \neq 0$. Two representations π_1, π_2 of A are called quasi-equivalent if $\pi_1(A)$ and $\pi_2(A)$ are normal faithful representations of each other. It is well-known that π_1 and π_2 are quasi-equivalent to each other if and only if $F_{\pi_1} = F_{\pi_2}$.

III.1. Definition:

Let (A,G,α) be a C*-dynamical system with G a topological group then:

a) A representation π on H is called covariant if there exists a unitary continuous representation U(g) of G on H with

$$U(g)\ \pi(x)\ U^*(g) = \pi(\alpha_g x).$$

b) A representation π is called quasi-covariant if there is a representation π_1 which is quasi-equivalent to π and which is at the same time a covariant representation.

We now want to generalize a result which is known for locally compact groups and the additional assumption that α_g acts strongly continuous on A.

III.1. Theorem

Let (A,G,α) be a C*-dynamical system with G a topological group and let π be a representation of A, then π is quasi-covariant if and only if

1. F_π is invariant under the action of α_g

2. α_g acts strongly continuous on F_π, this means for $\varepsilon > 0$ and $\omega \in F_\pi$ exist a neighbourhood N G of the identity such that

$$||\alpha_g \omega - \omega|| < \varepsilon \quad \text{for } g \in N.$$

Proof: The necessity of these conditions is well-known and can be found in (3). For proving the converse $l^1(G,A)$ be the set of functions $g \to A$ such that $\sum_G ||x(g)|| < \infty$. $l^1(G,A)$ becomes a normed *-algebra if we introduce as sum the pointwise addition and

$$\{\{x\}x\{y\}\}(h) = \sum_{g \in G} x(g)\alpha_g\, y(g^{-1}h),$$

$$\{x\}^*(g) = \alpha_g(g^{-1})^*.$$

There is a natural embedding of A into $l^1(G,A)$ given by

$$i(x) = \begin{cases} x \text{ for } g = 1 \\ 0 \text{ for } g \neq 1 \end{cases} \tag{1}$$

If A contains the identity then we also have an embedding of G into $l^1(G,A)$ given by

$$\rho_g(h) = \begin{cases} 1 \text{ for } h = g \\ 0 \text{ for } h \neq g \end{cases} \tag{2}$$

If A does not contain the identity then ρ_g lies in the multiplier algebra of A (see e.g. (10) section 3.1.2 for the definition of multipliers). Between these quantities exist the relation

$$\rho_g i\,(x)\,\rho_g{-1} = i\,(\alpha_g x). \tag{3}$$

If $\{x(g)\}$ is an element in $l^1(G,A)$ then it can be written as

$$\{x\} = \sum_{g \in G} i(x(g))\rho_g.$$

Next let $\omega \in S(A)$ we want to show by using the standard interpolating functionals defined in section II that there exists an extention $\tilde{\omega}$ of

ω to $l^1(G,A)$.

III.2. Definition:

Let (A,G,α) be a C*-dynamical system. For $\omega \in S(A)$ define

$$\phi_{g_1,g_2} = \phi_{\alpha_{g_1}\omega,\alpha_{g_2}\omega}$$

as in definition II.2. By $\tilde{\omega}$ denote the continuous linear functional on $l^1(G,A)$ which is given by the formular

$$\tilde{\omega}(\{x\}) = \sum_{g\in G} \phi_{1,g^{-1}}(x(g)).$$

III.3. Lemma:

Let (A,G,α) be a C*-dynamical system and ω a state of A, then

a) $\tilde{\omega}$ is a state of $l^1(G,A)$.

b) Let $(\pi_{\tilde{\omega}},H_{\tilde{\omega}})$ be the G.N.S. representation of $l^1(G,A)$ then

α) $\pi_{\tilde{\omega}}(i(x)) = \tilde{\pi}(x)$
defines a representation of A on $H_{\tilde{\omega}}$.

β) $\pi_{\tilde{\omega}}(\rho_g) = U(g)$
defines a unitary representation of G on $H_{\tilde{\omega}}$.

γ) $U(g)$ implements the automorphism α_g this means

$$U(g)\tilde{\pi}(x)\,U^*(g) = \tilde{\pi}(\alpha_g x)$$

c) The folium of $\tilde{\pi}$ is the smallest α_g invariant folium containing ω.

Proof: It is clear that $\tilde{\omega}$ defines a norm-continuous linear functional on $l^1(G,A)$. Let $\{x\} \in l^1(G,A)$ then we have

$$\{x\}^*\{x\}\ (h) = \sum_g \{\alpha_g(x(g^{-1}))^*\}\ \alpha_g(x(g^{-1}h)$$

and hence

$$\tilde{\omega}(\{x\}^*\{x\}) = \sum_{h,g} \phi_{1,h^{-1}} (\alpha_g x^*(g^{-1})x(g^{-1}h))$$

$$= \sum_{h,g} (\alpha_g \phi_{1,h^{-1}}) (x^*(g^{-1})x(g^{-1}h)$$

From the definition of $\phi_{g,h}$ and theorem II.1. c) follows:

$$\alpha_g \phi_{\omega,\alpha_{h^{-1}}\omega} = \phi_{\alpha_g\omega,\alpha_g(\alpha_{h^{-1}}\omega)}.$$

Since α_g applied to ω is the dual action follows $\alpha_g(\alpha_{h^{-1}}\omega) = \alpha_{h^{-1}g}\omega$. So we have $\alpha_g\phi_{1,h^{-1}} = \phi_{g,h^{-1}g}$. From this we get

$$\tilde{\omega}(\{x\}^*\{x\}) = \sum_{h,g} \phi_{g,h^{-1}g} (x^*(g^{-1})x(g^{-1}h)) \geq 0$$

by theorem II.1. b). This schows a). From equation (1) follows that $\tilde{\pi}(x)$ is a representation of A. From (2) follows U(g) is a representation of G. It remains to show that U(g) is unitary. Let $\psi_{\tilde{\omega}}$ be the cyclic vector of π_ω then we have

$$U(g)\pi_{\tilde{\omega}}(\{x\})\psi_{\tilde{\omega}} = \pi_{\tilde{\omega}}(\rho_g\{x\})\psi_{\tilde{\omega}}.$$

From the definition of ρ_g and equation (2) follows $\rho_g^* = \rho_{g^{-1}}$ and hence we obtain:

$$||U(g)\pi_{\tilde{\omega}}(\{x\})\psi_{\tilde{\omega}}||^2 = ||\pi_{\tilde{\omega}}(\rho_g\{x\})\psi_{\tilde{\omega}}||^2$$

$$= \tilde{\omega}([\rho_g\{x\}]^*\rho_g\{x\}) = \tilde{\omega}(\{x\}^*\rho_{g^{-1}}\rho_g\{x\})$$

$$= \tilde{\omega}(\{x\}^*\{x\}) = ||\pi_{\tilde{\omega}}(\{x\})\psi_{\tilde{\omega}}||^2 .$$

Since the vectors $\pi_{\tilde{\omega}}(\{x\})\psi_{\tilde{\omega}}$ are dense in $H_{\tilde{\omega}}$ follows that U(g) is unitary. The implementability equation follows from (3). This shows b). Since the elements of the form $i(x)\rho_g$ generate linearly the algebra $l^1(G,A)$ follows the set of vectors of the form $\tilde{\pi}(x)U(g)\psi_{\tilde{\omega}}$ is total in $H_{\tilde{\omega}}$

From $(\tilde{\pi}(x)U(g)\psi_{\tilde{\omega}}, \tilde{\pi}(y)\tilde{\pi}(x)U(g)\psi_{\tilde{\omega}}) = \omega(\alpha_{g^{-1}} x^*yx) = x^*\alpha_g\omega x(y)$

follows that the folium of $\tilde{\pi}$ is the folium generated by the set

$\{\alpha_g\omega, g \in G\}$. This proves the lemma.

Next we turn to the question of the continuity of the unitary representation $U(g)$.

III.4. Lemma:

With the same notations and assumptions as in lemma III.3. we get:

If the folium of $\tilde{\pi}_\omega$ is contained in A_c^* then follows $U(g)$ is a continuous representation of the group G.

Proof: Since $U(g)$ is unitary, it is sufficient to show that $U(g)$ is strongly continuous on a dense set of vectors in $H_{\tilde{\omega}}$. We take the set $\pi_{\tilde{\omega}}(\{x\})\psi_{\tilde{\omega}}$. Moreover since G is a group and $U(g)$ is unitary, we only need to show the continuity at $g = 1$.

$$||(U(h)\pi_{\tilde{\omega}}(\{x\}) - \pi_{\tilde{\omega}}(\{x\}))\psi_{\tilde{\omega}}||^2 =$$

$$\tilde{\omega}(\{x\}^*(U(h)-1)^* (U(h)-1)\{x\}) =$$

$$2\tilde{\omega}(\{x\}^*\{x\}) - 2\mathrm{Re}\ \tilde{\omega}(\{x\}^*\rho_h\{x\}).$$

Using definition III.2. for $\tilde{\omega}$ and equation (3) for the action ρ_h we obtain together with theorem II.1.c):

$$\tilde{\omega}(\{x\}^*\rho_h\{x\}) = \sum_{g,g'} \phi_{1,g'^{-1}}(\alpha_g[x^*(g^{-1})\alpha_h x(g^{-1}g'h)])$$

$$= \sum_{g,g'} \phi_{g,g'^{-1}g}(x^*(g^{-1})\alpha_h x(g^{-1}g'h)) = \sum_{g,g_1} \phi_{g,h^{-1}g_1^{-1}}(x^*(g^{-1})\alpha_h x(g_1)).$$

Where we have set $g^{-1}g'h^{-1} = g_1$. With the equation we have:

$$|\tilde{\omega}(\{x\}^*\{x\}) - \tilde{\omega}(\{x\}\rho_h\{x\})|$$

$$=|\sum_{g,g_1} \phi_{g,g_1^{-1}}(x^*(g^{-1})x(g_1) - \phi_{g,h^{-1}g_1^{-1}}(x^*(g^{-1})\alpha_h x(g_1)|$$

$$\leq|\sum_{g,g_1} (\phi_{g,g_1^{-1}} - \phi_{g,h^{-1}g_1^{-1}})\ (x^*(g^{-1})\alpha_h x(g_1))|$$

$$+ \left| \sum_{g,g_1} \phi_{g,g_1^{-1}} (x^*(g^{-1})(1 - \alpha_h)x(g_1)) \right|$$

For estimating the first term we use theorem II.1.d)

$$||\phi_{g,g_1^{-1}} - \phi_{g,h^{-1}g_1^{-1}}|| \leq ||(1 - \alpha_h)\ \alpha_{g_1^{-1}}\omega||.$$

Since $x(g_1)$ vanishes except on a finite set Γ and since $\alpha_{g_1^{-1}}\omega \in A_c^*$ exist a neighbourhood U_1 with $||(1 - \alpha_h)\alpha_{g_1^{-1}}\omega|| < \varepsilon$ for $h \in U_1$ and $g_1 \in \Gamma$. Hence the first term can be estimated by $\varepsilon||\{x\}||^2$ for $h \in U_1$. For the second term remark that lemma III.3. implies that $\phi_{g,g_1^{-1}}$ and hence $x^*(g^{-1})\phi_{g,g_1^{-1}}$ is a $\pi_{\tilde{\omega}}$ normal functional and hence belongs to A_c^* by assumption. So for $g_1 \in \Gamma$ exist a neighbourhood U_2 such that $||(1 - \alpha_h)x^*(g^{-1})\phi_{g,g_1^{-1}}|| < \varepsilon$ for $h \in U_2$. This means the second term is majorized by $\varepsilon_2 ||\{x\}||^2$. Therefore the whole expression is majorized by $2\varepsilon||\{x\}||^2$ for $h \in U_1\ \ U_2$. But this shows the conitnuity of $U(g)$.

End of the proof of theorem III.1.

If $\{\pi_\alpha, \alpha \in I\}$ are covariant representations (with $U_\alpha(g)$ continuous) then $\sum_{\alpha \in I} \pi_\alpha$ is again a covariant representation. Hence

$$\sum_{\omega \in F} \tilde{\pi}_\omega$$

is a covariant representation. From the assumptions of the theorem follows by lemma III.3. $F_{\tilde{\pi}_\omega}\ \ F$ and hence $F_{\Sigma\tilde{\pi}_\omega}\ \ F$. On the other hand $\omega \in F_{\tilde{\pi}_\omega}$ so that $F_{\Sigma\tilde{\pi}_\omega} = F$. This means $\sum_{\omega \in F} \tilde{\pi}_\omega$ is quasi-equivalent to the given representation which is therefore quasi-covariant.

IV. Final remarks

Since the covariant and quasi-covariant representations are governed by the space $M_{*,c}$ one would like to know the structure of this space. In (5) the following properties are derived:

(i) If $\phi \in M_{*,c}$ then also $|\phi| \in M_{*,c}$ where $|\phi|$ is the absolute value of ϕ defined by Tomita (13)

(ii) If H_c denotes the smallest Hilbert-space containing P_c then H_c is invariant under the modular involution.

(iii) The cone P_c is self-dual in H_c.

(iv) H_c is generated by P_c this means for every $\psi \in H_c$ exist four elements $\xi_i \in P_c$ with $\xi = \sum_{j=0}^{3} (i)^j \xi_j$

Very little is known about the characterization of covariant representations. The reason for this seems to be that one has to deal with two obstructions simultaniously. The one is the multiplicity theory. The other is the fact that one has to deal with projective representations of the group. Since one can construct from a projective group representation a representation of the group itself by passing to the direct product of the original Hilbert-space with itself these problems do not appear if we only ask for quasi-covariant representations.

References:

(1) Araki, H.: Some properties of modular conjugation operator of a von Neumann algebra and a non-commutative Radon-Nikodym theorem with a chain rule. Pac. J. Math. 50, 309-354 (1974)

(2) Araki, H.: Introduction to relative Hamiltonian and relative entropy. Preprint Marseille (1975)

(3) Borchers, H.-J.: On the implementability of automorphism groups. Comm. math. Phys. 14, 305-314 (1969)

(4) Borchers, H.-J.: Über C*-Algebren mit lokalkompakten Symmetriegruppen. Nachr. d. Akad. d. Wissensch. in Göttingen, Heft 1, 1-18 (1973)

(5) Borchers, H.-J.: C*-Algebras and Automorphism Groups - to be published in Comm. math. Phys.

(6) Bratteli, O. and D.W. Robinson: Operator Algebras and Quantum Statistical Mechanics I. Springer Verlag: New York, Heidelberg, Berlin (1979)

(7) Connes, A.: Sur le théorème de Radon Nikodym pour le poids normaux fidèles semifinis. Bull. Sci. Math. 2ème Sér. 97, 253-258 (1973)

(8) Connes, A.: Caractérisation des algèbres de von Neumann comme espaces vectoriels ordonés. Ann. Inst. Fourier 26, 121-155 (1974)

(9) Doplicher, S., D. Kastler and D.W. Robinson: Covariance algebras in field theory and statistical mechanics. Comm. math. Phys. 3, 1-28 (1966)

(10) Haagerup, U.: The standard form of von Neumann algebra. Math. Scand. 37, 271-283 (1975)

(11) Petersen, G.K.: C*-Algebras and their Automorphism Groups. Academic Press: London, New York, San Francisco (1979)

(12) Takesaki, M.: Tomita's Theory of Modular Hilbert Algebras and its Applications. Lecture Notes in Mathematics Vol. 128, Springer Verlag: Berlin, Heidelberg, New York (1970)

(13) Tomita, M.: Spectral theory of operator algebra I. Math. J. Okayama Univ. 9, 63-98 (1959)

(14) Tomita, M.: Quasi-standard von Neumann algebras. Mimeographed notes (1967)

JUMP PROCESSES AND APPLICATIONS TO THE TRIGONOMETRIC INTERACTION

Ph. COMBE*
R. RODRIGUEZ*
M. SIRUGUE
M. SIRUGUE-COLLIN**

Centre de Physique Théorique, CNRS, Marseille
*and Université d'Aix-Marseille II, Luminy, Marseille
**and Université de Provence, Marseille

INTRODUCTION

These last years, probabilistic methods in Quantum Physics gave rise to a great research activity (see for example [1] and the references therein), especially in Constructive Quantum Field Theory where diffusion processes play a central role. In that formulation, time is complexified in a first step, and the relativistic field theory is recovered from the Osterwalder Schrader reconstruction theorem ([2], [3]).

It was recently remarked [4], in the case of finitely degrees of freedom, and for real time, that jump type processes may be considered for the study of quantum dynamics.

Some papers [5][6][7] have been published in this approach. They are concerned with, in particular, the problem of the definition of the Feynman path integral for infinite anticommutative systems on lattices and self-interacting Boson fields, the interactions in any case being sufficiently smooth, in a sense we shall make precise later. Let us note that this is an alternative approach to the one proposed by S. Albeverio, R. Hoegh-Krohn, C. Dewitt-Morette, in the program of mathematical definition of the Feynman path integral. This later is defined by duality in terms of Fresnel integrals which is a bounded linear functional on some Banach space.

In this talk, we carry on the study of the role played by jump processes in Quantum evolution problems, for finite and infinite numbers of degrees of freedom.

In Section 2, we show how the Schrödinger equation in the interaction picture may be viewed as a backward Kolmogorov equation associated to jump type processes which satisfy some stochastic differential equations.

Such an approach leads [6] to a representation of quantum thermal functionals in terms of path integrals with respect to mixed diffusion and jump processes.

We also give a realization of the underlying probability space and the considered processes. This may be generalized to models of field theory we shall consider in Section 3.

There, we show how it is possible to build a path integral for the evolution operator of a Bose field, the self-interaction of which being of trigonometric type, with suitable space time and ultraviolet cut-off.

Thus, we give a result (see also [5]) on the problem of removing the ultraviolet cut-off for the Sine Gordon model. It is obtained directly in the real time sector, without passage to the Euclidean region by using the properties of functionals which have to be integrated with respect to Poisson probability measures.

II. PATH INTEGRALS IN THE INTERACTION PICTURE

2.1. STOCHASTIC PROCESSES OF POISSON TYPE

Let E be a 2n-dimensional real symplectic vector space, the phase space of a quantum mechanical system with n degrees of freedom and let $[0,T]$ be an interval of time.

Let μ be a positive bounded measure on E and ν the Poisson random measure on $[0,T] \times E$ ([10], [11]) such that

1. $\forall t \in [0T]$, $B \in \mathcal{B}(E)$, a Borel set of E, $\nu(t,B)$ is a Poisson process with parameter $\mu(B)$ that is

$$(2.1) \qquad E(\exp\{ i\lambda\, \nu(t,B)\}) = \exp\{ t(e^{i\lambda} - 1)\, \mu(B)\}$$

2. $\forall t \in [0T]$, $B_1, B_2 \in \mathcal{B}(E)$ two Borel sets of E such that $B_1 \cap B_2 = \emptyset$, thus $\nu(t,B_1)$ and $\nu(t, B_2)$ are two independent random variables.

We shall be interested, in the following, by the processes $Y_{vt}(s)$ defined as :

$$(2.2) \qquad Y_{vt}(s) = v + \int_t^s \int_E \alpha_{T-\tau}(u)\ \nu(d\tau, du)$$

where $s,t \in [0,T]$, $v \in E$ and $\alpha_\tau(u)$ is a solution of the Hamilton equations

$$(2.3) \qquad \frac{d}{d\tau} \alpha_\tau(u) = J \nabla h(\alpha_\tau(u)))$$

with initial condition $\alpha_0(u) = u$. Here J is the usual complex structure on the phase space and h is an at most quadratic classical hamiltonian.

The process $Y_{ot}(T)$ has its trajectories which end at $t = T$ on $v = 0$. Its characteristic functional is given by

$$(2.4) \qquad E(e^{i\langle \beta ,\, Y_{ot}(T)\rangle}) =$$

$$= \exp\left\{ (t-T)\, \mu(E) + \int_t^T dt \int_E d\mu(u) \exp i \langle \beta\ ,\ \alpha_{T-\tau}(u)\rangle \right\}$$

where $\langle\ ,\ \rangle$ is the scalar product in E. Furthermore, the process Y_{vt} will be defined if $\int_E |u|\, d\mu(u) < \infty$.

2.2. SCHRÖDINGER EQUATION IN THE INTERACTION PICTURE AS BACKWARD KOLMOGOROV EQUATION

Let us consider the real valued process $\eta_{\gamma t}(s)$ defined in terms of $Y_{vt}(s)$ by

$$(2.5) \qquad \eta_{\gamma t}(s) = \gamma + \int_t^s \int_E \left\{ \sigma(Y_{vt}(\tau), \alpha_{T-\tau}(u)) - \frac{\pi}{2} \right\} \nu(d\tau, du)$$

Here γ is real and σ is the symplectic structure on E given by

$$(2.6) \qquad \sigma(u,u') = \frac{1}{2}\langle u, Ju'\rangle$$

Now, if G is a C^2-function from $E \times R$ in C, then the following functions

$$(2.7) \qquad f(t,v,\gamma) = E(G(Y_{vt}(T), \eta_{\gamma t}(T))$$

which is such that

$$(2.8) \qquad \lim_{t \to T} f(t,v,\gamma) = G(v,\gamma)$$

satisfies the backward Kolmogorov equation [10]

$$(2.9) \qquad \frac{\partial f}{\partial t} + \int \left\{ f(t, v+\alpha_{T-t}(u), \gamma + \sigma(v, \alpha_{T-t}(u)) - \frac{\pi}{2}) - f(t,v,\gamma) \right\} d\mu(u) = 0$$

We shall show that the Schrödinger equation in the interaction picture is an equation like (2.9) for a special choice of the function G. This can be done by considering a Weyl system on E namely a mapping

$$u \longrightarrow W_u$$

such that

$$(2.10) \qquad W_u W_{u'} = \exp\ i\,\sigma(u,u')\, W_{u+u'}$$

$$(2.11) \qquad W_u^* = W_{-u}$$

Furthermore, the *-automorphism of the Weyl algebra defined by

$$(2.12) \qquad \hat{\alpha}_t(W_u) = W_{\alpha_t(u)}$$

where $\alpha_t(u)$ has been defined in (2.3), is unitarily implemented by a one-parameter group U_t^o, whose infinitesimal generator will be called H_o. Typically, H_o is $-\Delta$ or the harmonic oscillator hamiltonian.

Let us consider a perturbation of H_o by a potential V (which may depend on velocities)

$$(2.13) \qquad V = \int_E d\mu(u)\, W_u$$

where μ is the (positive) bounded measure considered previously. The more general case where μ is a complex measure can be handled along the same lines [6].

Let us consider the functional

$$(2.14) \qquad g(t,v,\phi,\psi) = (\phi,\ W_v \exp i\, H_o(T-t) \exp - i(H_o + V)(T-t)\psi)$$

Here ϕ and ψ are vectors in the Hilbert space of the Weyl representation of W_u. The function g satisfies

$$(2.15) \qquad \frac{\partial g}{\partial t} = i \int_E d\mu(u) \exp(i\sigma(v, \alpha_{T-t}(u)))\ g(t, v + \alpha_{T-t}(u))$$

with the condition

$$(2.16) \qquad \lim_{t \to T} g(t,v,\phi,\psi) = (\phi,\ W_v \psi)$$

which is a bounded function of v.

So, by choosing in (2.8) the function G such that

$$(2.17) \qquad G(v,\gamma) = (\phi,\ W_v \psi)\, e^{i\gamma}$$

one can write (2.9) as

$$(2.18) \qquad \frac{\partial f}{\partial t} + \int \left\{ f(t, v + \alpha_{T-t}(u), \gamma)\, e^{i\left\{-\frac{\pi}{2} + \sigma(v, \alpha_{T-t}(u))\right\}} - f(t,v,\gamma) \right\} d\mu(u) = 0$$

The last term may be eliminated by considering $\{\exp(T-t)\,\mu(E)\}\, f(t,v,\gamma)$ instead of $f(t,v,\gamma)$ and the Schrödinger equation in the interaction picture is recovered by getting $\gamma = 0$. So, one has

$$(2.19) \qquad g(t,v,\phi,\psi) = e^{(T-t)\mu(E)}\ E((\phi,\ W_{Y_{vt}(T)}\ \psi)\ e^{i\,\eta_{ot}(T)})$$

The first part of the functional to be integrated in that expression is completely determined as soon as one takes a representation of the Weyl algebra. We shall concentrate on the second part of the functional which has an interesting meaning especially if one thinks of generalization to the case of infinite numbers of degrees of freedom.

2.3. PATH INTEGRALS IN PHASE SPACE

The formulae (2.19) is of the type of Feynman integral in phase space. We shall detail the form of the process $\eta_{ot}(T)$ which appears in it. In a first step, one has

$$(2.20)\qquad \eta_{ot}(T) = \int_t^T \int_E \sigma(Y_\tau, \alpha_{T-\tau}(u))\ \nu(d\tau, du) + \int_t^T \int_E \sigma(v, \alpha_{T-\tau}(u))\ \nu(d\tau, du) - \frac{\pi}{2} N_t$$

where $Y_{ot}(\tau) = Y_\tau$ and N_t is the usual Poisson process with parameter $\mu(E)$, unit jumps and which ends at $t = T$.

The first part of the process, $\eta'_{ot}(T)$ can be written

$$(2.21)\qquad \eta'_{ot}(T) = \int_t^T \sigma(Y_\tau\ , dY_\tau)$$

The second part is a boundary one and may be written $\eta^2_{ot}(T) = \int_t^T \sigma(v, dY_\tau)$.
In order to give a more illustrative picture of $\eta'_{ot}(T)$, let us specialize ourselves to the more interesting case, for our purpose, of Y_τ. This corresponds to an α_t as defined in (2.3) associated to the quantum harmonic oscillator hamiltonian

$$(2.22)\qquad H_o = \sum_{i=1}^{N} \frac{1}{2}\ (P_i^2 + \chi^2\ Q_i^2)$$

In that case, the process $\eta^1_{ot}(T)$ takes the following form

$$(2.23)\qquad \eta'_{ot}(T) = -\int_t^T\int_t^T\int_E\int_E \Big\{\Delta_R(\tau'-\tau)\ p.p' + \Delta'_R(\tau'-\tau)(p.q' - p'.q) + \chi^2\Delta_R(\tau'-\tau)\ q.q'\Big\}\ \nu(d\tau, dqdp)\ \nu(d\tau', dq'dp')$$

where Δ_R (resp. Δ'_R) is the retarded function (resp. its derivative) defined as

$$(2.24)\qquad \Delta_R(t) = \frac{\sin\chi t}{\chi}\ \theta\ (t)$$

In (2.23), the point . stands for the scalar product in R^N. In any case, the formula (2.19) for the quantum functional $g(t,v,\phi,\psi)$ may be considered as a path integral in phase space.

2.4. A REALIZATION OF THE UNDERLYING PROBABILITY SPACE

It is convenient, especially for the extension to the case of an infinite number of degrees of freedom, to give an explicit realization of the underlying probability space as well as the processes which have been considered (see also [5]).

Let T be a positive time. $\Omega \equiv \Omega(R^{2n}, [0,T])$ is the disjoint union

(2.25) $$\Omega = \bigsqcup_{n=0} \Omega_n$$

where $\Omega_n = \{ w = (n, t_1 \ldots t_n, u_1, \ldots, u_n)$, the t_i's being in increasing order $0 \leq t_1 < t_2, \ldots, < t_n \leq T$ and $u_i \in R^{2n}$. Furthermore, $\Omega_0 = \{w_0\}$.

Let $\mathcal{F}$ be the Borel σ-algebra generated by the following open sets

(2.26) $$\mathcal{V}^n_{\mathcal{a}_i \mathcal{B}_i} = \{ w = (n, t_i, u_i) \quad t_i \in \mathcal{a}_i \quad u_i \in \mathcal{B}_i \}$$

where the $\mathcal{a}_i$'s are disjoint ordered Lebesgue measurable subsets of $[0,T]$ and the $\mathcal{B}_i$'s are Borel subsets of R^{2n}.

Thus, given a bounded positive measure μ on R^{2n}, one can construct a positive bounded measure on Ω as the unique measure which extends the additive set of functions defined on the $\mathcal{V}^n$'s by

(2.27) $$P(\mathcal{V}^n_{\mathcal{a}_i \mathcal{B}_i}) = \prod_{i=1}^{n} |\mathcal{a}_i| \, \mu(\mathcal{B}_i)$$

where $|\mathcal{a}_i|$ is the Lebesgue measure of $\mathcal{a}_i$.

Now, if G is a measurable functional its expectation value is given by

(2.28) $$E(F) = \sum_{n \geq 0} \int_0^T dt_n \ldots \int_0^{t_2} dt_1 \int_E d\mu(u_1), \ldots \int_E d\mu(u_n)$$
$$G(w = (n, t_1, \ldots, t_n, u_1, \ldots, u_n))$$

The processes we have considered before then take explicit forms which may be convenient in some applications. For example, the process

(2.29) $$Z_t(w = (n, t_1, \ldots, t_n, u_1, \ldots, u_n)) = \sum_{i=1}^{n} \alpha_{t-t_i}(u_i) \; \theta(t_i - t)$$

where θ is the step function, is a realization of

(2.30) $$Z_t(s) = \int_t^s \int_E \alpha_{t-\tau}(u) \; \nu(d\tau, du)$$

for $s = T$. The usual Poisson process with parameter $\mu(E)$ is obtained through $\alpha \equiv 1$, the identity automorphism of E .

III. GENERALIZATION AND APPLICATION TO RELATIVISTIC BOSE FIELD THEORY

3.1. THE CASE OF AN INFINITE NUMBER OF DEGREES OF FREEDOM

We shall consider a neutral scalar Bose field and perturbations of the free field which can be written

(3.1) $$V = \lambda \int_{\mathcal{H}} d\mu(f) \, : e^{i\phi(f)} :$$

where $\mathcal{H} = \mathcal{S}_{R}(R^s)$ is the Schwartz space of real functions, μ is a bounded positive measure on $\mathcal{H}$, $\phi(f)$ is the tested field, : : is some ordering.

In order to construct the (generalized) Poisson process suitable for what follows, let us consider the mapping

(3.2) $$f \in \mathcal{S}_{R}(R^s) \longrightarrow C(f)$$
$$= \exp\left\{|\lambda| \int_0^T dt \int_{\mathcal{H}} d\mu(g) \left\{ e^{i \iint d\xi \, d\zeta \, \Delta_R^m(\xi-\zeta\, ; \, t) \, g(\xi) \, f(\zeta)} - 1 \right\}\right\}$$

where Δ_R^m is the usual retarded function, which is defined as the Fourier transform of

(3.3) $$\Delta_R^m(k,t) = \frac{\sin \sqrt{k^2 + m^2}\, t}{\sqrt{k^2 + m^2}} \quad \theta(t)$$

m being the mass of the field, θ the step function.
The mapping is normalized, continuous and of positive type. So it is the characteristic functional of a generalized process X_R

(3.4) $$C(f) = E(e^{i X_R(f)})$$

In fact, we can realize this process in the following way. Let $\mathcal{H}$ be equipped with the discrete topology in order to avoid topological problems and let $\Omega \equiv \Omega(\mathcal{H}, [0,T])$ be the disjoint union

(3.5) $$\Omega = \bigsqcup_{n \geq 0} \Omega_n$$

where $\Omega_0 = \{\eta_0\}$, that is, Ω_0 is reduced to one point

(3.6) $$\Omega_n = \left\{ \eta = (n, t_1, \ldots, t_n, g_1, \ldots, g_n) \; n \in N \quad 0 \leq t_1 < t_2, \ldots, < t_n \leq T \quad g_i \in \mathcal{H} \right\}$$

As before, one can construct on Ω a bounded measure P_μ in such a way that, for any measurable functional F, its expectation value $E(F)$ is given by

$$(3.7) \qquad E(F) = \int_\Omega P_\mu(d\eta)\, F(\eta)$$

$$= \sum_{n \geq 0} \int_0^T dt_n \int_0^{t_n} dt_{n-1} \cdots \int_0^{t_2} dt_1 \int d\mu(g_n), \ldots, \int d\mu(g_1)$$

$$F(\eta = (n, t_1, \ldots, t_n, g_1, \ldots, g_n))$$

So, the following generalized Poisson process X_R (random field) as considered in (3.4), has the form $X_R \equiv X_{R,t=0}$, where

$$(3.8) \qquad X_{R,t}(x)(\eta = (n, t_i, g_i)) = \sum_{i=1}^{n} \int_{R^s} d\xi\, \Delta_R^m(x - \xi\, ,\, t_i - t)\, g_i(\xi)$$

Now, let H_0 denote the generator of the free evolution which exists in suitable representation of the algebra of canonical commutation relations of the field. We shall only consider here the Fock representation , the Fock vacuum being denoted ψ_0.

Thus, (see [5]), one can write a formula completely analogous to the one obtained in the interaction picture of ordinary quantum mechanics, namely

$$(3.9) \qquad (\psi_0, \exp iTH_0 \exp -iT(H_0+V)\, \psi_0)$$

$$= \int_\Omega P_\mu(d\eta) \left\{ e^{-\frac{i}{4} \int_0^T dt \int_{R^s} dx\, X_{F,t}(x)(\Box + m^2)\, X_{F,t}(x)}\; e^{-\frac{i\pi}{2} N_0} \right\} (\eta)$$

where $N_0 \equiv N_{t=0}$ is the usual Poisson process with parameter $\mu(\mathcal{H}) < \infty$ and which ends at $t = T$.

Here $X_{F,t}$ is the process defined in the same way as $X_{R,t}$ in (3.8) with Δ_R^m replaced by Δ_F^m, the Feynman propagator. The meaning of the first part of the functional to be integrated is

$$(3.10) \qquad \left\{ \int_0^T dt \int_{R^s} dx\, X_{F,t}(x)\, (\Box + m^2)\, X_{F,t}(x) \right\} (\eta = (n, t_1, \ldots, t_n, g_1, \ldots, g_n)$$

$$= -\, 2i \sum_{i=1}^{n} \int_{R^s} X_{F,t_i}(x)\, g_i(x)\, dx$$

Moreover, in the definition of V, the Wick ordering has been defined with respect to Δ_F^m.

In what follows, we shall consider as an example of application, a more specific model of self-interacting Bose field, namely the trigonometric interaction.

3.2. EXAMPLE : THE TRIGONOMETRIC INTERACTION

In that section, we shall restrict ourselves to fields with ultraviolet cut-offs in a space time box $\Lambda \times [0T]$. Here Λ is a cube in R^s.

The trigonometric self-interaction of the neutral scalar Bose field is defined in the following way

$$(3.11) \qquad H_I^\kappa = \lambda \int_\Lambda dx \int_R d\nu(\alpha) : e^{i\alpha \phi(\chi_\kappa^x)} :_\delta$$

χ_k^x is the function

$$(3.12) \qquad \chi_\kappa^x(y) = \frac{1}{\kappa^s} \chi\left(\frac{1}{\kappa} (x-y)\right)$$

$\kappa \in R$, where χ is an ultraviolet cut-off function of $\mathcal{D}(R^s)$ which is symmetric, positive such that $\chi(x) = 0$ if $|x_i| \geqslant 1$, $i = 1,2,\dots,s$ and such that $\int_{R^s} \chi(x)\, dx = 1$. $\phi(\chi_k^x)$ will be written $\phi_k(x)$ as usual.

ν is a bounded measure on R such that $\overline{\nu}(\alpha) = \nu(-\alpha)$, in order to have self-adjoint interactions. Moreover, $:\ :_\delta$ is the Wick ordering with respect to Δ_F^δ.

We shall call $P_{\Lambda|\lambda|}$ the Poisson measure in that case. It is defined in the following way.

Let us consider the σ-Borel algebra of open sets $\vartheta^n_{\{a_i \mathcal{B}_i \mathcal{C}_i\}}$ of $\Omega \equiv \Omega(\Lambda \times (R, \mathcal{B}(R), \nu), [0,T])$ where $\mathcal{B}(R) \equiv \mathrm{Borel}(R)$, which are defined as

$$(3.13) \qquad \vartheta^0 = \{\eta_0\}$$

$$(3.14) \qquad \vartheta^n_{\{a_i \mathcal{B}_i \mathcal{C}_i\}_{i=1,2,\dots,n}} = \left\{ \eta = (n, t_1,\dots,t_n, x_1,\dots,x_n, \alpha_1,\dots,\alpha_n) \quad t_i \in a_i \ \ x_i \in \mathcal{B}_i \ \ \alpha_i \in \mathcal{C}_i \right\}$$

where the a_i's are disjoint ordered Lebesgue measurable subsets of $[0,T]$, the $\mathcal{B}_i$'s are Lebesgue measurable subsets of Λ, and $\mathcal{C}_i \in (R, \mathcal{B}(R), \nu)$, $i=1,2,\dots,n$ $\forall n$.

Then

$$(3.15)\qquad P_{\wedge|\lambda|}\,(\vartheta^{n}_{\mathcal{A}_i\mathcal{B}_i\mathcal{C}_i}) = |\lambda|^{n}\prod_{i=1}^{n}|\mathcal{A}_i|\,|\mathcal{B}_i|\,\nu(\mathcal{C}_i)$$

$|\mathcal{A}_i|$ (resp. $|\mathcal{B}_i|$) being the Lebesgue measure of $\mathcal{A}_i$ (resp. $\mathcal{B}_i$). Thus, $P_{\wedge|\lambda|}$ extends to a bounded measure on Ω, the Poisson measure.

Now, let us consider coherent states ψ_{fg}, $(f,g) \in (\mathcal{S}_R(R^s) \times \mathcal{S}_R(R^s))$, in the Fock representation, namely

$$(3.16)\qquad \psi_{fg} = \exp i(\pi(f) - \phi(g))\,\psi_o$$

where ψ_o is the Fock vacuum, and $\phi(g)$, $\pi(f)$ are the conjugate field operators.

One can thus prove, using the preceding results, the following

Proposition (3.17)

Let ψ_{fg}, $f,g \in \mathcal{S}_R(R^s)$ be coherent states in the Fock representations. Let H_o be the relativistic free hamiltonian for the Boson scalar neutral field with ultraviolet cut-off in $(s+1)$ space time dimensions. Then

$$(3.18)\qquad (\psi_{fg}\ \exp i\,H_oT\ \exp -i(H_o + H_I^K)T\ \ \psi_{fg}) = \int_{\Omega} P_{\wedge|\lambda|}\,(d\eta)\ \Xi^{fg}_{km}(\eta)$$

Here Ξ^{fg}_{km} is given by

$$(3.19)\qquad \Xi^{fg}_{km}(\eta = (n, t_i, x_i, \alpha_i)) = \mathcal{E}^{fg}_{km}(\eta)\ \ \mathcal{F}_{km}(\eta)$$

where

$$(3.20)\qquad \mathcal{E}^{fg}_{km}(\eta) = \exp\Big\{\sum_{i=1}^{n} i\,\alpha_i \iint d\xi\, d\zeta\ \chi_k(\xi)\Delta^m_R(x_i - \xi - \zeta,\ t_i)\ g(\zeta)\Big\}$$

$$\exp\Big\{-i\sum_{i=1}^{n}\alpha_i \iint d\xi\, d\zeta\ \chi_k(\xi)\ \partial_o\Delta^m_R(x_i - \xi - \zeta,\ t_i)\ f(\zeta)\Big\}$$

$$(3.21)\qquad \mathcal{F}_{km}(\eta) = \exp\Big\{\frac{1}{4}\sum_{i=1}^{n}\alpha_i^2\int_R |\tilde{\chi}_\kappa(k)|^2((k^2+\delta^2)^{-\frac{1}{2}}dk - (k^2+m^2)^{-\frac{1}{2}})$$

$$\exp\Big\{-\frac{1}{4}\sum_{i\neq j}^{n}\alpha_i\alpha_j \iint d\xi\, d\zeta\ \ \chi_\kappa(x_i - \xi)\Delta^m_F(\xi - \zeta,\ t_i - t_j)\ \ \chi_\kappa(x_j - \zeta)\Big\}$$

In (3.21), $\tilde{\chi}$ is the Fourier transform of χ.

The problem of removing the ultraviolet cut-off can be handled in some special cases using (3.18). The advantage of that formulation is that it works in the real time sector, more precisely without passage to the Euclidean region, [8], [9]. Moreover, evolution matrix elements have been written as well defined functional integrals with respect to bounded Poisson measures.

3.3. THE ZERO MASS SINE GORDON MODEL WITHOUT ULTRAVIOLET CUT-OFF

This model corresponds to the case where ν is a measure such that

$$\nu = \frac{1}{2}(\delta_{\alpha} + \delta_{-\alpha}) \tag{3.22}$$

where δ is the Dirac measure. In order to control the singularities which appear in (3.19), (3.20), (3.21), we have considered the following case : s = 1, m = 0, and f = 0.

More precisely, in a first step, one can show :

$$\lim_{m \to 0} \int_{\Omega} P_{\Lambda|\lambda|}(d\eta)\, \Xi^{g}_{km}(\eta) = \int_{\Omega} P_{\Lambda|\lambda'|}(d\eta)\, \Xi^{g}_{k}(\eta) \tag{3.23}$$

with $\Xi^{og}_{km} \equiv \Xi^{g}_{km}$. Here, a typical element in $\Omega \equiv \Omega(\Lambda \times Z_2, [0,T])$ has the following structure

$$\eta = (n\ ,\ t_1, \ldots, t_n,\ x_1, \ldots, x_n,\ \varepsilon_1, \ldots, \varepsilon_n)$$

where n is an integer, $0 \leqslant t_1 < t_2 \ldots < t_n \leqslant T$, $x_i \in \Lambda$ and $\varepsilon_i = \pm 1$.

In (3.23), one has

$$\lambda' = \lambda \left(\frac{2}{\delta}\, e^{-\gamma}\right)^{\frac{\alpha^2}{4\pi}} \tag{3.24}$$

where γ is the Euler constant. Moreover, one has

$$\Xi^{g}_{k}(\eta) = \xi^{g}_{k}(\eta)\, \mathcal{F}_{k}(\eta) \tag{3.25}$$

where $\xi^{g}_{k}(\eta)$ is the pointwise limit of $\xi^{g}_{km}(\eta)$ as $m \to 0$ and $\mathcal{F}_k(\eta)$ has the following form

$$\mathcal{F}_k(\eta) = \left(-\frac{i\lambda'}{|\lambda'|}\right)^{n} \exp\left\{\frac{n\alpha^2}{4}\int dk \left(|\chi_{\kappa}(k)|^2 - \frac{1}{2\pi}\right)\left((k^2+\delta^2)^{-\frac{1}{2}} - |k|^{-1}\right)\right\}$$
$$\exp\left\{\frac{\alpha^2}{4\pi}\sum_{i\neq j}\varepsilon_i\,\varepsilon_j \iint d\xi\, d\zeta\, \chi_{\kappa}(x_i-\xi)\, C_F(\xi-\zeta, t_i-t_j)\, \chi_{\kappa}(x_j-\zeta)\right\} \tag{3.26}$$

If $\sum_{i=1}^{n} \varepsilon_i = 0$

(3.27) $\mathcal{F}_{\kappa}(\eta) = 0$ if $\sum_{i=1}^{n} \varepsilon_i \neq 0$

Furthermore

(3.28) $$C_F(x,t) = \frac{1}{2} \ln(t^2-x^2) + \frac{i\pi}{2} \quad \text{if} \quad |x| < |t|$$

$$= \frac{1}{2} \ln(x^2-t^2) \quad \text{if} \quad |t| > |x|$$

Actually, one can write

(3.29) $$\mathcal{F}_{\kappa m}(\eta) = \mathcal{F}_{\kappa}(\eta)\ m^{\frac{\alpha^2}{4\pi}(\sum_{i=1}^{n} \varepsilon_i)^2}\ G_{\kappa m}(\eta)\ \left(\frac{\lambda'}{\lambda}\right)^n$$

with $G_{\kappa m}(\eta) \longrightarrow 1$ pointwise as $m \longrightarrow 0$. The Lebesgue dominated convergence theorem working here, one has the result (3.23).

Now, the problem of removing the ultraviolet cut-off, that is $\kappa \longrightarrow 0$, can be solved along the same lines as in Euclidean field theory (see e.g. [8]), the integrals being taken here with respect to the Poisson measure.

So, one can prove ([5]).

For $\alpha^2 < 2\pi$, the following limit exists

(3.30) $$\lim_{\kappa \to 0} \int_{\Omega} P_{\Lambda|\lambda'|}(d\eta)\, \Xi_{\kappa}^{g}(\eta) = \int_{\Omega} P_{\Lambda|\lambda'|}(d\eta)\, \Xi^{g}(\eta)$$

where

(3.31) $$\Xi^{g}(\eta) = \mathcal{E}^{g}(\eta)\, \mathcal{F}(\eta)$$

$\mathcal{E}^{g}(\eta)$ is the pointwise limit of $\mathcal{E}^{g}_{k}(\eta)$ as $k \to 0$, namely

(3.32) $$\mathcal{E}^{g}(\eta) = \exp\left\{ i\alpha \sum_{i=1}^{n} \varepsilon_i \int_R d\zeta\, \Delta_R^0(x_i - \zeta\ ,\ t_i)\, g(\zeta) \right\}$$

and $\mathcal{F}(\eta)$ is such that

(3.33) $$\mathcal{F}(\eta) = \left(-\frac{i\lambda'}{|\lambda'|}\right)^n \exp\left\{ -\frac{\alpha^2}{4\pi} \sum_{i \neq j} \varepsilon_i\, \varepsilon_j\, C_F(x_i - x_j, t_i - t_j) \right\} \text{ if } \sum_{i=1}^{n} \varepsilon_i = 0$$

(3.34) $$\mathcal{F}(\eta) = 0 \text{ if } \sum_{i=1}^{n} \varepsilon_i \neq 0$$

Thus, one can state the

Theorem (3.35)

Let H_I^K be the Sine Gordon self-interaction of a scalar neutral relativistic Bose field of mass m with ultraviolet cut-off, in a one-dimensional space box Λ, namely

$$H_I^K = \lambda \int_\Lambda dx : \cos\alpha\, \phi_K(x) :_\delta \tag{3.36}$$

where λ, α are real, δ is a strictly positive constant and the Wick ordering is taken with respect to Δ_F^δ. Let H_o be the free relativistic Hamiltonian and let ψ_g be coherent states in the Fock representation (for the mass m), $g \in \mathcal{S}_R(R)$. Then, for $\alpha^2 < 2\pi$, the following limit exists as a well defined path integral

$$\lim_{K\to 0}\lim_{m\to 0} (\psi_g \exp iTH_o \exp -iT(H_o + H_I^K)\ \psi_g) = \int_\Omega P_{\Lambda|\lambda'|}\ (d\eta)\, \mathcal{I}^g(\eta) \tag{3.37}$$

$\mathcal{I}^g(\eta)$ (resp. λ') is defined in (3.31) - (3.34) (resp. (3.24)).

- REFERENCES -

[1] J. GLIMM, A. JAFFE, Quantum Physics, Springer Verlag (1981).

[2] K. OSTERWALDER, in Constructive Quantum Field Theory, G. Velo and A.S. Wightman Eds., Springer Verlag (1973).

[3] B. SIMON, The $P(\phi_2)$-Euclidean (Quantum) Field Theory, Princeton Series in Physics, Princeton University Press (1974).

[4] V.P. MASLOV, A.M. CHEBOTAREV, Theoretical and Mathematical Physics, vol. 28, 3, 793-805 (1976).

[5] Ph. COMBE, R. HOEGH-KRØHN, R. RODRIGUEZ, M. SIRUGUE and M. SIRUGUE-COLLIN,
i) - Poisson Processes on Groups and Feynman Path Integrals, Commun.Math.Phys. 77, 269-288 (1980).

ii)- Feynman Path Integrals and Poisson Processes with Piecewise Classical Paths, Journ. Math. Phys. 23, 405-411 (1982).

iii)- Zero Mass, 2-Dimensional Real Time Sine Gordon Model Without Ultraviolet Cut-Off, to appear in Ann. Inst. H. Poincaré.

[6] Ph. COMBE, R. RODRIGUEZ, M. SIRUGUE and M. SIRUGUE-COLLIN,
i) - High Temperature Behaviour of Quantum Mechanical Thermal Functionals, Preprint Marseille, CPT-82/P.1366.

ii)- Definition de l'Intégrale de Feynman et Processus à Sauts, Preprint Marseille, CPT-82/PE.1392, to be published in the Proceedings of RCP 25, Strasbourg (1982).

[7] S. ALBEVERIO, Ph. BLANCHARD, Ph. COMBE, R. HOEGH-KRØHN, M. SIRUGUE, Local Relativistic Invariant Flows for Quantum Fields, Preprint Marseille CPT-81/ P. 1342.

[8] J. FRÖHLICH, Commun.Math.Phys. 47, 233 (1976).

[9] J. FRÖHLICH, Renormalization Theory, International School of Mathematical Physics, G. Velo, A.S. Wightman Eds.

[10] I.I. GIKHMAN, A.V. SKOROKHOD, The Theory of Stóchastic Processes I, II, Springer Verlag (1974).

[11] Y.V. PROHOROV, Y.A. ROZANOV, Probability Theory, Springer Verlag (1969).

GENERALIZED HOMOMORPHISMS BETWEEN C*-ALGEBRAS AND KK-THEORY

Joachim CUNTZ*
Department of Mathematics
University of Pennsylvania
Philadelphia, PA 19104

In the category of C*-algebras, the natural morphisms are, of course, the C*-algebra homomorphisms, i.e. the algebra homomorphisms which commute with the involution (such homomorphisms are automatically norm-decreasing). One disadvantage of this notion of a morphism is that, given two C*-algebras A and B, there are in general very few homomorphisms from A to B; in fact, typically, often the only possible homomorphism is the 0-map. For the study of homotopy and topological properties of C*-algebras one would like a more flexible class of morphisms. In practice, very different looking C*-algebras can be homotopy equivalent, in a generalized sense, but the equivalence is induced, rather than by a homomorphism, by what we shall call a quasihomomorphism. A quasihomomorphism from A to B is, essentially, a pair of homomorphisms $\phi,\bar{\phi}$ from A to E, where E is a C*-algebra containing an ideal J with $J \subset B$, such that $\phi(x) - \bar{\phi}(x) \in J$ for all $x \in A$. With this definition, there are, in general, many quasihomomorphisms from A to $K \otimes B$, where K is the algebra of compact operators on a Hilbert space of countably infinite dimension. Another advantage is, that the "negative" $(\bar{\phi},\phi)$ of a quasihomomorphism $(\phi,\bar{\phi})$ is again a quasihomomorphism.

Our aim, in this article, is to develop some of the fundamental properties of Kasparov's KK-theory [4] on the basis of the notion of a quasihomomorphism. The group $KK(A,B)$ can be defined as the set of all homotopy classes of quasihomomorphisms from A to $K\otimes B$. We then construct the product $KK(A_0,A_1) \times KK(A_1,A_2) \to KK(A_0,A_2)$ which is the heart of Kasparov's theory (and generalizes the composition of homotopy classes of homomorphisms). Our definitions allow us to avoid many of the technicalities that are part of Kasparov's work, like graded algebras, Hilbert modules or stabilization. We even, as a variation, replace Kasparov's main technical theorem by a related theorem of Pedersen, in the construction of the product. The associativity of the product, a point of great importance, which is rather mysterious in [4], and still somewhat unnatural in [6], is nearly automatic in our approach. Additional

*partially supported by NSF

minor features are, that we do not use the rather unintuitive stabilization theorem at all, and that we don't have to assume the existence of strictly positive elements in A_1 and A_2 for the construction of the product.

For certain applications, our frame for KK-theory may prove less clumsy and easier to handle than the one of Kasparov. Our approach emphasizes the nature of elements of KK as generalized homomorphisms, cf. also [2], while in Kasparov's work an element of KK appears rather as a generalized elliptic operator. Thus, in situations as in [1] where one handles elements of KK arising naturally from such operators, it will, presumably, be preferable to work with Kasparov's definitions and with the product as defined in [4] or the modification in [1].

This article is meant to be read in conjunction with the expository article [3] which gives an outline of K-theory for C*-algebras and contains a discussion and some applications of the present ideas.

I am very grateful to Daniel Kastler and his colleagues at the CPT-CNRS, Marseille-Luminy for their hospitality during a stay which gave me the leisure to get acquainted with Kasparov's theory.

1. Preliminaries.

An important tool in C*-algebra theory is the so-called multiplier algebra $M(A)$ for a C*-algebra A. A multiplier (or double centralizer) of A is a pair (T_1,T_2) where T_1,T_2 are (necessarily bounded) linear maps from A to A such that $T_1(xy) = xT_1(y)$, $T_2(xy) = T_2(x)y$ and $T_1(x)y = xT_2(y)$ for all $x,y \in A$. The set of all multipliers of A, with the obvious addition, the product $(T_1,T_2)(T_1',T_2') = (T_1 \cdot T_1', T_2' \cdot T_2)$ the involution $(T_1,T_2)^* = (T_2^*,T_1^*)$ where $T_i^*(x) = (T_i(x^*))^*$, $i = 1,2$, and the operator norm, forms a C*-algebra. This C*-algebra is called the multiplier algebra of A and denoted $M(A)$. If A is faithfully represented on a Hilbert space H , then $M(A)$ is isomorphic to the algebra $\underline{A} = \{x \in L(H) \mid xa, ax \in A$ for all $a \in A\}$ via the isomorphism $\underline{A} \ni x \to (T_1,T_2) \in M(A)$ with $T_1(a) = ax$, $T_2(a) = xa$ for $a \in A$; cf. [5,3.12]. The C*-algebra A is contained in $M(A)$ as an ideal (all our ideals are closed). We have the following basic fact.

1.1 Proposition. Let J be an ideal in a C*-algebra E and $\phi\colon J \to B$ a homomorphism into some C*-algebra B. Then ϕ extends uniquely to a homomorphism $\phi'\colon E \to M(\phi(J))$. If J is essential (i.e. $xJ = \{0\}$ implies $x = 0$ for $x \in E$) and ϕ is injective, then ϕ' is injective.

Proof. The characterization of multipliers as double centralizers immediately shows that ϕ' , defined by $\phi'(x)\phi(y) = \phi(xy)$, or more formally, $\phi'(x) = (T_1,T_2)$ with $T_1(\phi(y)) = \phi(yx)$, $T_2(\phi(y)) = \phi(xy)$, is a well defined homomorphism (note that $\mathrm{Ker}\ \phi \subset J$ is an ideal in E). The second assertion is then obvious. q.e.d.

In particular, this proposition shows that, when J is an essential ideal in E , then E may be considered, in a natural way, as a subalgebra of $M(J)$.

For later reference, we state here Kasparov's main technical theorem, which he uses for the construction of the product, in its simplest form.

1.2 Theorem. (cf. [4,§3,Theorem 3]). Let J admit a countable approximate unit and let $R_1,R_2,\mathcal{D}$ be separable subalgebras of $M(J)$ such that $R_1R_2 \subset J$ and such that $\mathcal{D}$ derives R_1 and R_2 , i.e. the sets of commutators $[\mathcal{D},R_i]$ are contained in R_i , $i = 1,2$. Then there exist positive elements M,N in $M(J)$ such that $M + N = 1$, $MR_1,NR_2 \subset J$, and such that $[M,\mathcal{D}],[N,\mathcal{D}] \subset J$.

The proof of this theorem is rather complicated. We replace it in our approach by the following (related and not much simpler) theorem which is due to G.K. Pedersen.

1.3 Theorem. (cf. [5,8.6.15]). Let $\pi: B \to A$ be a surjective homomorphism between separable C*-algebras. If $\dot{\delta}$ is a derivation of A , then there is a derivation δ of B such that $\dot{\delta}\pi = \pi\delta$.

By a derivation of B (or A) we mean here a bounded *-derivation, i.e. an everywhere defined linear map $\delta: B \to B$ such that $\delta(xy) = \delta(x)y + x\delta(y)$ and $\delta(x^*) = \delta(x)^*$ for all $x,y \in B$.

2. Quasihomomorphisms and KK .

Let us begin with a definition.

2.1 Definition. Let A and B be C*-algebras. A prequasihomomorphism from A to B is a triple $(\phi,\overline{\phi},\mu)$ where $\phi,\overline{\phi}$ are homomorphisms from A to a C*-algebra E, which contains an ideal J, such that $\phi(x) - \overline{\phi}(x) \in J$ for all $x \in J$, and μ is a homomorphism from J to B.

We write, symbolically, $(\phi,\overline{\phi}): A \to E \triangleright J \overset{\mu}{\to} B$. Every ordinary homomorphism $\alpha: A \to B$ gives, of course, a prequasihomomorphism (α,o,id) with $E = J = B$. Moreover, ordinary homomorphisms can be composed with prequasihomomorphisms. If $(\phi,\overline{\phi},\mu)$ is as above and $\alpha: A' \to A$, $\beta: B \to B'$ are homomorphisms, then $(\phi,\overline{\phi},\mu)\alpha = (\phi\alpha,\overline{\phi}\alpha,\mu)$ is a prequasihomomorphism from A' to B, while $\beta(\phi,\overline{\phi},\mu) = (\phi,\overline{\phi},\beta\mu)$ is a prequasihomomorphism from A to B'.

With every prequasihomomorphism $\Phi = (\phi,\overline{\phi},\mu)$ from A to B we associate a linear map $D_\Phi: A \to B$ and a bilinear map $Q_\Phi: A \times A \to B$ setting $D_\Phi(x) = \mu(\phi(x)-\overline{\phi}(x))$ and $Q_\Phi(x,y) = \mu(\overline{\phi}(x)(\phi(y)-\overline{\phi}(y)))$ for $x,y \in A$. These two maps contain all the information on Φ which is relevant for our purposes.

2.2 Definition. A quasihomomorphism from A to B is a prequasihomomorphism $(\phi,\overline{\phi}): A \to E \triangleright J \overset{\mu}{\to} B$, where E is generated, as a C*-algebra, by $\phi(A)$ and $\overline{\phi}(A)$, J is generated, as an ideal in E, by $\phi(x) - \overline{\phi}(x)$, $x \in A$, J is essential in E and, finally, μ is an embedding.

If μ is an embedding we most often omit it in our notation and simply write $(\phi,\overline{\phi})$ for $(\phi,\overline{\phi},\mu)$.

2.3 Proposition. (a) If Φ_1 and Φ_2 are quasihomomorphisms from A to B such that $D_{\Phi_1} = D_{\Phi_2}$ and $Q_{\Phi_1} = Q_{\Phi_2}$, then $\Phi_1 = \Phi_2$. (b) For every prequasihomomorphism Φ from A to B, there is a unique quasihomomorphism Φ' from A to B such that $D_\Phi = D_{\Phi'}$ and $Q_\Phi = Q_{\Phi'}$.

Proof. (a) Let Φ_i, $i = 1,2$ be given by $(\phi_i,\overline{\phi}_i): A \to E_i \triangleright J_i \subset B$. Since J_i is the C*-algebra generated by $D_{\Phi_i}(x)$ and $Q_{\Phi_i}(x,y)$, $x,y \in A$ for $i = 1,2$, it follows that $J_1 = J_2 = J$ as a subalgebra of B. Moreover, since J_i is essential we may consider E_i as a subalgebra of $M(J)$. The equality $Q_{\Phi_1} = Q_{\Phi_2}$ then shows that $\overline{\phi}_1(x)j = \overline{\phi}_2(x)j$

for all $j \in J$, whence $\overline{\phi}_1 = \overline{\phi}_2$. Since $D_{\Phi_1} = D_{\Phi_2}$ it follows that also $\phi_1 = \phi_2$ which implies $E_1 = E_2$. (b) To obtain Φ' from Φ given by $(\phi,\overline{\phi}): A \to E \triangleright J \overset{\mu}{\to} B$, simply replace J by the C*-subalgebra of B generated by $D_\Phi(x)$, $Q_\Phi(x,y)$ $x,y \in A$, and E by the image in $M(J)$, of the C*-algebra generated by $\phi(A)$ and $\overline{\phi}(A)$, under μ', where μ': $E \to M(J)$ is the canonical extensions of μ, c.f. 1.1. The uniqueness of Φ' follows from (a). q.e.d.

If Φ is a quasihomomorphism from A to B and α: $A' \to A$, β: $B \to B'$ are homomorphisms we still write, by abuse of notation, $\Phi\alpha$, $\beta\Phi$ for the quasihomomorphisms defined by the prequasihomomorphisms $\Phi\alpha$ and $\beta\Phi$. Two quasihomomorphisms Φ and Ψ from A to B are called homotopic, if there is a family P_t, $t \in [0,1]$ of quasihomomorphisms from A to B such that the maps $t \to D_{P_t}(x)$ and $t \to Q_{P_t}(x,y)$ are continuous for each $x,y \in A$, and such that $P_0 = \Phi$, $P_1 = \Psi$.

2.4 <u>Proposition</u>. Two quasihomomorphisms Φ and Ψ as above are homotopic if and only if there exists a quasihomomorphism P from A to $B([0,1])$, the C*-algebra of continuous B-valued functions on $[0,1]$, such that $q_0P = \Phi$ and $q_1P = \Psi$, where q_t: $B([0,1]) \to B$ for $t \in [0,1]$ are the evaluation maps.

<u>Proof</u>. If P: $A \to B([0,1])$ is a quasihomomorphism then q_tP is a homotopy connecting q_0P and q_1P. Conversely, let $(\alpha_t,\overline{\alpha}_t): A \to E_t \triangleright J_t \subset B$ be a homotopy connecting Φ to Ψ. Set $\hat{E} = \bigoplus_{t\in[0,1]} E_t$ and define $\alpha,\overline{\alpha}$: $A \to \hat{E}$ by $\alpha(x) = \oplus\alpha_t(x)$ and $\overline{\alpha}(x) = \oplus\overline{\alpha}_t(x)$. Let E be the C*-algebra generated by $\alpha(A)$ and $\overline{\alpha}(A)$ in $\hat{E}$ and J the ideal generated in E by the elements $\alpha(x) - \overline{\alpha}(x)$, $x \in A$. Then, because of the continuity assumptions $J \subset B([0,1])$ and $P = (\alpha,\overline{\alpha}): A \to E \triangleright J \subset B([0,1])$ is a quasihomomorphism such that $q_tP = (\alpha_t,\overline{\alpha}_t)$ for all $t \in [0,1]$. In particular $q_0P = \Phi$ and $q_1P = \Psi$. q.e.d.

Recall that K denotes the C*-algebra of compact operators on a Hilbert space of countably infinite dimension. One has $K\otimes K \cong K$ and there is a natural embedding of $K\oplus K$ in $M_2(K) \cong K$. We use this embedding to define the direct sum of two quasihomomorphisms $(\phi_1,\overline{\phi}_1)$ and $(\phi_2,\overline{\phi}_2)$ from A to $K\otimes B$ by

$$(\phi_1\oplus\phi_2,\overline{\phi}_1\oplus\overline{\phi}_2): A \to E_1\oplus E_2 \triangleright J_1\oplus J_2 \subset K\otimes B \oplus K\otimes B \subset K\otimes B .$$

This direct sum is a prequasihomomorphism from A to $K \otimes B$.

We now define $KK(A,B)$ as the set of homotopy classes of quasihomomorphisms from A to B. We write $[\phi,\bar{\phi}]$ for the homotopy class of $(\phi,\bar{\phi})$. With the addition $[\phi_1,\bar{\phi}_1] + [\phi_2,\bar{\phi}_2] = [\phi_1\oplus\phi_2,\bar{\phi}_1\oplus\bar{\phi}_2]$ this is an abelian group. In fact, the inverse for $[\phi,\bar{\phi}]$ is $[\bar{\phi},\phi]$ – $(\phi\oplus\bar{\phi},\bar{\phi}\oplus\phi)$ is easily seen to be homotopic to 0 by rotating $\bar{\phi}\oplus\phi$ to $\phi\oplus\bar{\phi}$. Moreover, KK is a contravariant functor in the first variable and a covariant functor in the second variable. Every homomorphism $\alpha: A' \to A$ induces a homomorphism $\alpha^*: KK(A,B) \to KK(A',B)$ by $\alpha^*[\phi,\bar{\phi}] = [(\phi,\bar{\phi})\alpha]$ while a homomorphism $\beta: B \to B'$ induces a homomorphism $\beta_*: KK(A,B) \to KK(A,B')$ by $\beta_*[\phi,\bar{\phi}] = [\beta(\phi,\bar{\phi})]$.

Remark 1. It is very easy to see that for every C*-algebra B one has $KK(\mathbb{C},B) = K(B)$ where K is the ordinary K-functor (cf. e.g. [3] for its definition). In fact, one of the possible definitions of $K(B)$ is as the group of homotopy classes of pairs of projections $(p,\bar{p})$ in $K\otimes\tilde{B}$ ($\tilde{B} = B$ with unit adjoined) such that $p - \bar{p} \in K \otimes B$. Clearly, every such pair of projections defines a prequasihomomorphism $(\phi,\bar{\phi})$ from $\mathbb{C}$ to $K \otimes B$, by $\phi(1) = p$, $\bar{\phi}(1) = \bar{p}$. This gives a map from $K(A)$ to $KK(\mathbb{C},A)$. Using the stabilization theorem one sees that this map is an isomorphism.

Remark 2. Higher functors $KK_n(A,B)$, $n > 0$ can be defined as the n-th homotopy group of the set of quasihomomorphisms from A to B. One has Bott periodicity $KK_{n+2}(A,B) \cong KK_n(A,B)$.

3. The product.

We are now going to construct the product $KK(A_0,A_1) \times KK(A_1,A_2) \to KK(A_0,A_2)$ assuming that A_0 is separable. Given quasihomomorphisms $(\alpha,\bar{\alpha}): A_0 \to E_1 \rhd J_1 \subset K\otimes A_1$ and $(\beta,\bar{\beta}): A_1 \to E_2 \rhd J_2 \subset K\otimes A_2$, we want to define their composition. First of all, we can of course extend $(\beta,\bar{\beta})$ to $(id_K\otimes\beta, id_K\otimes\bar{\beta}): K\otimes A_1 \to K\otimes E_2 \rhd K\otimes J_2 \subset K\otimes K\otimes A_2 = K\otimes A_2$. We denote this extended quasihomomorphism from $K\otimes A_1$ to $K\otimes A_2$ still by $(\beta,\bar{\beta})$. It obviously defines, by restriction, a quasihomomorphism $(\beta,\bar{\beta})|_{J_1}: J_1 \to E_2' \rhd J_2' \subset K\otimes A_2$. Now we would like to define the individual compositions $\beta\alpha, \bar{\beta}\alpha, \beta\bar{\alpha}, \bar{\beta}\bar{\alpha}$. This leads to the concept of an extendible quasihomomorphism.

We say that a quasihomomorphism $(\phi,\bar{\phi}): J_1 \to \mathcal{D} \rhd J \subset B$ is <u>extendible</u> to $E_1 \rhd J_1$, if ϕ and $\bar{\phi}$ both extend to homomorphisms from E_1 into some C*-algebra $\hat{\mathcal{D}}$ containing $\mathcal{D}$ such that

$$\phi(e)x = \lim_\lambda \phi(e)\phi(u_\lambda)x$$

$$\bar{\phi}(e)x = \lim_\lambda \bar{\phi}(e)\bar{\phi}(u_\lambda)x$$

$e \in E_1$, $x \in \mathcal{D}$, u_λ an approximate unit for J_1 (this implies that we may choose $\hat{\mathcal{D}}$ as a subalgebra of $M(\mathcal{D})$).

3.1 <u>Proposition</u>. For every quasihomomorphism $(\phi,\bar{\phi}): J_1 \to \mathcal{D} \rhd J \subset K\otimes A_2$, there is a quasihomomorphism $(\phi^e,\bar{\phi}^e)$ from J_1 to $K\otimes A_2$ which is homotopic to $(\phi,\bar{\phi})$ and extendible to E_1. This quasihomomorphism is unique, up to homotopy in the category of quasihomomorphisms which are extendible to E_1.

<u>Proof</u>. Let K denote the C*-subalgebra of $M_2(\mathcal{D})$ generated by matrices of the form

$$\begin{pmatrix} \phi(x_1) & \phi(x_2)\bar{\phi}(y_2) \\ \bar{\phi}(y_3)\phi(x_3) & \bar{\phi}(x_4) \end{pmatrix} \qquad x_i, y_i \in J_1$$

and by all products of the form $a\,x\,b$, where a,b are such matrices and $x \in M_2(J)$. One has obvious embeddings of $M(\phi(J_1))$ and $M(\bar{\phi}(J_1))$ into $M(K)$ as $\begin{pmatrix} M(\phi(J_1)) & 0 \\ 0 & 0 \end{pmatrix}$ and $\begin{pmatrix} 0 & 0 \\ 0 & M(\bar{\phi}(J_1)) \end{pmatrix}$.

The homomorphisms $\begin{pmatrix} \phi & 0 \\ 0 & 0 \end{pmatrix}$ and $\begin{pmatrix} 0 & 0 \\ 0 & \bar{\phi} \end{pmatrix}$ therefore extend to homomorphisms from $M(J_1)$ to $M(K)$.

With $K_0 = K \cap M_2(J)$, the quotient K/K_0 is isomorphic to $M_2(\phi(J_1)/J)$ and K is separable. By Theorem 1.3, there exists therefore an automorphism $\sigma = e^{\delta}$ of K, δ a derivation, which lifts the automorphism $\mathrm{Ad}\begin{pmatrix} 0 & 1 \\ 1 & 0 \end{pmatrix}$ of K/K_0.

The pair $\phi^e = \begin{pmatrix} \phi & 0 \\ 0 & 0 \end{pmatrix}$ and $\overline{\phi}^e = \sigma\begin{pmatrix} 0 & 0 \\ 0 & \overline{\phi} \end{pmatrix}$ clearly defines a (pre) quasihomomorphism with the required properties (with $\hat{\mathcal{D}} \subset M(K))$. It is clear that the homotopy class of $(\phi^e, \overline{\phi}^e)$ does not depend on the choice of σ. If $\sigma' = e^{\delta'}$ is another lift, then $\sigma_t = e^{\delta' t} e^{\delta(1-t)}$, $t \in [0,1]$ is a continuous path of lifts connecting σ to σ'. Moreover if $(\phi, \overline{\phi})$ is already extendible to E_1, let K' be the C*-algebra generated by K and by the C*-algebra analogous to K with J_1 replaced by E_1, i.e. the algebra generated by matrices (x_{ij}) with $x_{11} \in \phi(E_1), x_{12} \in \phi(E_1)\overline{\phi}(E_1), x_{21} \in \overline{\phi}(E_1)\phi(E_1), x_{22} \in \overline{\phi}(E_1)$. Then K' contains K as an essential ideal, thus σ extends uniquely to K'. Moreover, the embedding of K in $M_2(\mathcal{D})$ extends to an embedding $j\colon K' \to M(M_2(\mathcal{D}))$ defined by $j(m)x = \lim_{\lambda} m(\hat{u}_\lambda x)$ for $m \in K'$, $x \in M_2(\mathcal{D})$, where $\hat{u}_\lambda = \begin{pmatrix} \phi(u_\lambda) & 0 \\ 0 & \overline{\phi}(u_\lambda) \end{pmatrix}$, u_λ an approximate unit for J_1.

The continuous path

$\mathrm{Ad}(F^{\pi i t}) \begin{pmatrix} 0 & 0 \\ 0 & \phi \end{pmatrix}$, $j e^{t\delta}\begin{pmatrix} 0 & 0 \\ 0 & \overline{\phi} \end{pmatrix}$ with $F = \begin{pmatrix} 0 & 1 \\ 1 & 0 \end{pmatrix}$ consisting of extendible prequasihomomorphisms from J_1 to $K \otimes A_2$ connects $(\phi, \overline{\phi})$ to $(\phi^e, \overline{\phi}^e)$.

Finally, if $(\psi, \overline{\psi})$ is a quasihomomorphism from J_1 to $K \otimes A_2$ which is homotopic to $(\phi, \overline{\phi})$ through a quasihomomorphism $(\rho, \overline{\rho})$ from J_1 to $K \otimes A_2([0,1])$, then $(\rho^e, \overline{\rho}^e)$ is a homotopy, in the category of extendible quasihomomorphisms, between $(\phi^e, \overline{\phi}^e)$ and $(\psi^e, \overline{\psi}^e)$. q.e.d.

<u>Remark</u>. Let us say that a quasihomomorphism $(\phi, \overline{\phi})$ as in 3.1, is in standard form, if it extends to $E_1 = M(J_1)$. This is the case if and only if $\phi(u_\lambda)$ and $\overline{\phi}(u_\lambda)$ converge (in the multiplier topology) to projections P and $\overline{P}$ in $M(\mathcal{D})$. Proposition 3.1 shows that every $(\phi, \overline{\phi})$ is homotopic to an, up to homotopy unique, quasihomomorphism in standard form (provided that J_1 is separable). The algebra K used above is then $\hat{P} M_2(\mathcal{D}) \hat{P}$, where $\hat{P} = P \oplus \overline{P} = \lim \hat{u}_\lambda$ and we have an embedding $M(K) \to M(M_2(\mathcal{D}))$. If, for instance $J_1 = E_1$, then $(\phi, \overline{\phi})$ automatically "extends" to E_1, without necessarily being in standard form.

In the following, we denote the extendible quasihomomorphism constructed from $(\beta, \overline{\beta})|_{J_1}$ by $(\beta^e, \overline{\beta}^e)\colon J_1 \to E_2'' \rhd J_2'' \subset K \otimes A_2$.

We can now form the compositions $\beta^e \alpha$, $\overline{\beta}^e \alpha$, $\beta^e \overline{\alpha}$, $\overline{\beta}^e \overline{\alpha}$. Write

$$D_1(x) = \beta^e\alpha(x) - \bar{\beta}^e\alpha(x)$$
$$\bar{D}_1(x) = \beta^e\bar{\alpha}(x) - \bar{\beta}^e\bar{\alpha}(x)$$
$$D_2(x) = \beta^e\alpha(x) - \beta^e\bar{\alpha}(x)$$
$$\bar{D}_2(x) = \bar{\beta}^e\alpha(x) - \bar{\beta}^e\bar{\alpha}(x)$$

for $x \in A_0$. Let A be the C*-algebra generated by all 2×2 matrices with entries in $D_i(A_0)$, $\bar{D}_i(A_0)$, $i = 1,2$ and by all matrices of the form

$$\begin{pmatrix} \beta^e\alpha(x) & 0 \\ 0 & \beta^e\alpha(x) \end{pmatrix} \quad x \in A_0 .$$

3.2 Proposition. Let $(\alpha,\bar{\alpha})$, $(\beta,\bar{\beta})$ and A be as above. There is an automorphism $\tau = e^\delta$ of A, δ a derivation of A, such that

$$\begin{pmatrix} \beta^e\alpha(x) & 0 \\ 0 & \bar{\beta}^e\bar{\alpha}(x) \end{pmatrix} - \tau\begin{pmatrix} \bar{\beta}^e\alpha(x) & 0 \\ 0 & \beta^e\bar{\alpha}(x) \end{pmatrix} \in M_2(J_2") \quad \text{for all } x \in A_0.$$

Proof. Let R denote the ideal $A \cap M_2(J_2")$ in A. Because of the identities

$$D_1(x) - \bar{D}_1(x) = D_2(x) - \bar{D}_2(x) = (\beta^e - \bar{\beta}^e)(\alpha - \bar{\alpha})(x)$$
$$D_1(x)D_2(y) = \beta^e(\alpha(x)(\alpha(y) - \bar{\alpha}(y))) - \bar{\beta}^e(\alpha(x)(\alpha(y) - \bar{\alpha}(y)))$$
$$+ D_1(y) - \bar{D}_1(y)$$

one has $D_1(x) - \bar{D}_1(x)$, $D_2(x) - \bar{D}_2(x)$, $D_1(x)D_2(y) \in R$ for all $x,y \in A_0$.

Let R_i be the ideal generated in A by $M_2(D_i(A_0))$, $i = 1,2$. From $D_i(xy) + D_i(x)D_i(y) = \beta^e\alpha(x)D_i(y) + D_i(x)\beta^e\alpha(x)$ for $i = 1,2$ and $D_1(A_0)D_2(A_0) \subset R$ we deduce that $R_1R_2 \subset R$. Thus the images $\dot{R}_1, \dot{R}_2$ of R_1, R_2 in $\dot{A} = A/R$ are orthogonal, i.e. $\dot{R}_1\dot{R}_2 = 0$, and there is an automorphism $\dot{\tau}$ of $\dot{A}$, of the form $\dot{\tau} = e^{\dot{\delta}}$ which acts like the identity on $\dot{R}_2$, like $\mathrm{Ad}\begin{pmatrix} 0 & 1 \\ 1 & 0 \end{pmatrix}$ on $\dot{R}_1$ and which leaves the image of $\begin{pmatrix} \beta^e\alpha(x) & 0 \\ 0 & \beta^e\alpha(x) \end{pmatrix}$ in $\dot{A}$ fixed for each $x \in A_0$.

For this $\dot{\tau}$ we have

$$\begin{pmatrix} \beta^e\alpha(x) & 0 \\ 0 & \bar{\beta}^e\bar{\alpha}(x) \end{pmatrix}^{\cdot} - \dot{\tau}\begin{pmatrix} \bar{\beta}^e\alpha(x) & 0 \\ 0 & \beta^e\bar{\alpha}(x) \end{pmatrix}^{\cdot} =$$

$$\begin{pmatrix} \beta^e\alpha(x) & 0 \\ 0 & (\beta^e\alpha(x) - D_1(x) - \bar{D}_2(x)) \end{pmatrix}^{\cdot} - \dot{\tau}\begin{pmatrix} (\beta^e\alpha(x) - D_1(x)) & 0 \\ 0 & (\beta^e\alpha(x) - D_2(x)) \end{pmatrix}^{\cdot}$$

$$= 0$$

for all $x \in A_0$ (note that $D_2(x)^{\bullet} = \overline{D}_2(x)^{\bullet}$) .

For τ we now take the lift for $\dot{\tau}$, whose existence is guaranteed by Theorem 1.3. q.e.d.

Remark. Using Kasparov's technical theorem (see 1.2) one gets, alternatively, τ of the form $\tau = \mathrm{Ad}\, F$ with

$$F = \begin{pmatrix} \sqrt{M} & \sqrt{N} \\ \sqrt{N} & -\sqrt{M} \end{pmatrix}$$

where $M,N \in M(J_2'')$ are positive, $M + N = 1$, $MD_1(A_0) \subset J_2''$, $ND_2(A_0) \subset J_2''$ M,N commute modulo J_2'' with $\beta^e\alpha(x)$ for all $x \in A_0$.

Write $(\beta\alpha,\overline{\beta\alpha})_\tau$ for the quasihomomorphism

$$\left(\begin{pmatrix} \beta^e\alpha & 0 \\ 0 & \overline{\beta}^e\overline{\alpha} \end{pmatrix} , \tau\begin{pmatrix} \overline{\beta}^e\alpha & 0 \\ 0 & \beta^e\overline{\alpha} \end{pmatrix}\right)$$

Clearly, the homotopy class of this quasihomomorphism does not depend on the choice of τ . If $\tau' = e^{\delta'}$ is another automorphism of A satisfying the condition in Proposition 3.2, then $(\beta\alpha,\overline{\beta\alpha})_{\tau_t}$ with $\tau_t = e^{\delta' t}e^{\delta(1-t)}$, $t \in [0,1]$, gives a homotopy between $(\beta\alpha,\overline{\beta\alpha})_\tau$ and $(\beta\alpha,\overline{\beta\alpha})_{\tau'}$. We will, therefore, omit the subscript τ.

3.3 Theorem. The map

$$([\alpha,\overline{\alpha}] \ , \ [\beta,\overline{\beta}]) \to [\beta\alpha,\overline{\beta\alpha}]$$

gives a well defined bilinear product

$$KK(A_0,A_1) \times KK(A_1,A_2) \to KK(A_0,A_2) \ .$$

Proof. We only have to check that the product is well-defined. If the quasihomomorphism $(\phi,\overline{\phi})$ from A_1 to $K\otimes A_2([0,1])$ implements a homotopy between $(\beta,\overline{\beta})$ and $(\beta',\overline{\beta}')$, the "product" $(\phi\alpha,\overline{\phi\alpha})$ gives a homotopy connecting $(\beta\alpha,\overline{\beta\alpha})$ to $(\beta'\alpha,\overline{\beta'\alpha})$. At the same time one sees that the product does not depend on the choice of the extendible quasihomomorphism $(\beta^e,\overline{\beta}^e)$ for $(\beta,\overline{\beta})$. Similarly, if $(\alpha,\overline{\alpha})$ is homotopic to $(\alpha',\overline{\alpha}')$ through a quasihomomorphism $(\psi,\overline{\psi})$ from A_0 to $K\otimes A_1([0,1])$, then $(\tilde{\beta}\psi,\overline{\tilde{\beta}\psi})$ gives a homotopy from $(\beta\alpha,\overline{\beta\alpha})$ to $(\beta\alpha',\overline{\beta\alpha'})$, if $(\tilde{\beta},\tilde{\overline{\beta}})$ is the quasihomomorphism $(\beta\otimes\mathrm{id},\overline{\beta}\otimes\mathrm{id})$ from $A_1\otimes C([0,1])$ to $(K\otimes A_2) \otimes C([0,1])$. q.e.d.

In (partial) agreement with Kasparov's notation we write $[\alpha,\bar{\alpha}][\beta,\bar{\beta}]$, rather than $[\beta,\bar{\beta}][\alpha,\bar{\alpha}]$, for the product $[\beta\alpha,\overline{\beta\alpha}]$.

3.4 Proposition. The product is functorial in the following sense. If $\phi: A_0' \to A_0$ and $\psi: A_2 \to A_2'$ are homomorphisms then

$$\phi^*([\alpha,\bar{\alpha}][\beta,\bar{\beta}]) = \phi^*([\alpha,\bar{\alpha}])[\beta,\bar{\beta}]$$

$$\psi_*([\alpha,\bar{\alpha}][\beta,\bar{\beta}]) = [\alpha,\bar{\alpha}]\psi_*([\beta,\bar{\beta}])$$

Proof. Obvious.

This "functoriality" is a special case of the associativity of the product proved in section 4.

4. Associativity.

We prove the associativity of the product by a sequence of simple observations.

4.1 Lemma. Let $(\alpha,\bar{\alpha}): A \to E \rhd J$ be a prequasihomomorphism from A to J , and $(\beta,\bar{\beta})$ a quasihomomorphism from J to B , which is the restriction of a quasihomomorphism $(\beta',\bar{\beta}')$ from E to B . Then the product of $[\alpha,\bar{\alpha}] \in KK(A,J)$ and $[\beta,\bar{\beta}] \in KK(J,B)$ is represented by the prequasihomomorphism $(\beta'\alpha\oplus\bar{\beta}'\bar{\alpha},\bar{\beta}'\alpha\oplus\beta'\bar{\alpha})$ from A to $M_2(B)$.

Proof. By definition, the product of $[\alpha,\bar{\alpha}] \in KK(A,E)$ and $[\beta',\bar{\beta}']$ $\in KK(E,B)$ is equal to this product. However, as a quasihomomorphism from A to E , $(\alpha,\bar{\alpha})$ is homotopic to $(\alpha\oplus 0,0\oplus\bar{\alpha}) = (\alpha,0) \oplus (0,\bar{\alpha})$ whence the lemma. q.e.d.

4.2 Lemma. With every quasihomomorphism $(\alpha,\bar{\alpha}): A \to E \rhd J \subset B$ we can associate a prequasihomomorphism $(\alpha',\bar{\alpha}'): A \to E' \rhd J \subset B$ which has all the properties of a quasihomomorphism, except that J is not necessarily essential, such that

(a) The quasihomomorphism defined by $(\alpha',\bar{\alpha}')$ is $(\alpha,\bar{\alpha})$.

(b) There is a split exact sequence $0 \to J \to E' \overset{q}{\to} A \to 0$ with right inverse for q given by α' , $q\alpha' = \mathrm{id}_A$.

Proof. Let $\alpha',\bar{\alpha}'$ be the homomorphisms $\mathrm{id} \oplus \alpha$, $\mathrm{id} \oplus \bar{\alpha}$ from A to $A \oplus E$ and E' the C*-algebra generated by $\alpha'(A)$, $\bar{\alpha}'(A)$. q.e.d.

4.3 Proposition. Let $0 \to J \to E \underset{\overset{\phi}{\curvearrowleft}}{\overset{q}{\to}} A \to 0$ be a split exact sequence of C*-algebras where E is separable. Then every quasihomomorphism $(\alpha,\bar{\alpha})$ from J to another C*-algebra B is homotopic to the restriction of a quasihomomorphism $(\rho,\bar{\rho})$ from E to $K\otimes B$.

Proof. The pair $(\mathrm{id},\phi q): E \to E \triangleright J$ defines a prequasihomomorphism from E to J. Then $(\rho,\bar{\rho})$ will be the product of $(\alpha,\bar{\alpha})$ and the quasihomomorphism $(\mathrm{id}^\cdot,\phi q^\cdot): E \to \dot{E} \triangleright J$ induced by $(\mathrm{id},\phi q)$ ($\dot{E}$ is the image of E in $M(J)$). This is the pair

$$\begin{pmatrix} \alpha^e \mathrm{id}^\cdot & 0 \\ 0 & \bar{\alpha}^e(\phi q^\cdot) \end{pmatrix}, \tau \begin{pmatrix} \bar{\alpha}^e \mathrm{id}^\cdot & 0 \\ 0 & \alpha^e(\phi q^\cdot) \end{pmatrix}$$

where τ is as in 3.2. The restriction of this pair to J is

$$\begin{pmatrix} \alpha^e & 0 \\ 0 & 0 \end{pmatrix}, \tau \begin{pmatrix} \bar{\alpha}^e & 0 \\ 0 & 0 \end{pmatrix}$$

which is clearly homotopic to $(\alpha,\bar{\alpha})$. q.e.d.

We are now ready to prove the following fundamental theorem.

4.4 Theorem. Let $(\alpha_1,\bar{\alpha}_1): A_0 \to K\otimes A_1$, $(\alpha_2,\bar{\alpha}_2): A_1 \to K\otimes A_2$, $(\alpha_3,\bar{\alpha}_3): A_2 \to K\otimes A_3$ be quasihomomorphisms, where A_0 is separable. Then $[\alpha_1,\bar{\alpha}_1]([\alpha_2,\bar{\alpha}_2][\alpha_3,\bar{\alpha}_3]) = ([\alpha_1,\bar{\alpha}_1][\alpha_2,\bar{\alpha}_2])[\alpha_3,\bar{\alpha}_3]$.

Proof. We represent $(\alpha_1,\bar{\alpha}_1)$ by a prequasihomomorphism $(\alpha_1',\bar{\alpha}_1'):$ $A_0 \to E_1' \triangleright J_1 \subset K\otimes A_1$ where E_1' is split and separable, as in 4.2. By 4.3 we may assume that $(\alpha_2,\bar{\alpha}_2)|_{J_1}$ is the restriction of a quasihomomorphism $(\rho_2,\bar{\rho}_2)$ from E_1' to $K\otimes A_2$ which we represent again by $(\rho_2',\bar{\rho}_2'): E_1' \to E_2' \triangleright J_2 \subset K\otimes A_2$ with E_2' split and separable. In the same way, we assume that $(\alpha_3,\bar{\alpha}_3)|_{J_2}$ is the restriction of a quasihomomorphism $(\rho_3,\bar{\rho}_3): E_2' \to E_3 \triangleright J_3 \subset K\otimes A_3$. Now, by Lemma 4.1, both triple products of the (α_i,α_i), $i = 1,2,3$, are represented simply by the prequasihomomorphism

$$(\rho_3\rho_2'\alpha_1' \oplus \bar{\rho}_3\bar{\rho}_2'\alpha_1' \oplus \rho_3\bar{\rho}_2'\bar{\alpha}_1' \oplus \bar{\rho}_3\rho_2'\bar{\alpha}_1',$$
$$\bar{\rho}_3\rho_2'\alpha_1' \oplus \rho_3\bar{\rho}_2'\alpha_1' \oplus \bar{\rho}_3\bar{\rho}_2'\bar{\alpha}_1' \oplus \rho_3\rho_2'\bar{\alpha}_1')$$

from A_0 to $M_4(J_3) \subset K\otimes A_3$. q.e.d.

5. Equivalence of our definitions with the ones of Kasparov.

Given two C*-algebras A and B, Kasparov constructs the group $KK(A,B)$ essentially in the following way. First, he defines $E(A,B)$ as the set of all pairs (Φ,F) where Φ is a homomorphism from A to $M_2(M(K\otimes B))$ of the form

$$\Phi(x) = \begin{pmatrix} \phi(x) & 0 \\ 0 & \phi'(x) \end{pmatrix}$$

and F is an operator in $M_2(M(K\otimes B)$ of the form

$$F = \begin{pmatrix} 0 & F_0 \\ F_0^* & 0 \end{pmatrix}$$

with $[F,\Phi(x)]$, $(F^2-1)\Phi(x) \in M_2(K\otimes B)$ for all $x \in A$. These last conditions are equivalent to $(\phi(x) - F_0\phi'(x)F_0^*)$, $(F_0F_0^* - 1)\phi(x)$, $F_0^*F_0 - 1)\phi'(x) \in K\otimes B$ for all $x \in A$.

A homotopy between (Φ,F) and (Φ',F') is, by definition, an element in $E(A,B([0,1]))$, whose restrictions to 0 and 1 give (Φ,F) and (Φ',F'), respectively. The group $KK(A,B)$ is then defined as the set of homotopy classes of elements in $E(A,B)$ with addition induced by the direct sum (it is unnecessary to divide $KK(A,B)$ in addition by homotopy classes of degenerate elements, as Kasparov does, cf. [6]).

Every quasihomomorphism $(\phi,\bar{\phi}) : A \to E \triangleright J \subset K\otimes B$ from A to B gives an element in $E(A,B)$, if we set

$$\Phi = \begin{pmatrix} \phi & 0 \\ 0 & \bar{\phi} \end{pmatrix} \qquad F = \begin{pmatrix} 0 & 1 \\ 1 & 0 \end{pmatrix}$$

Here, we embed J in $K\otimes B$ via the isomorphism

$$K\otimes B \cong \begin{pmatrix} \overline{J(K\otimes B)J} & \overline{J(K\otimes B)} \\ \overline{(K\otimes B)J} & K\otimes B \end{pmatrix}$$

given by the stabilization theorem (assuming that B has a strictly positive element). The embedding $J \to K\otimes B$, in the upper left corner, defined that way, extends to a homomorphism $E \to M(K\otimes B)$ which is an embedding, if J is essential in E.

Conversely, if $(\Phi,F) \in E(A,B)$ is normalized in the sense that $(F^2-1)\Phi(x) = 0$, $x \in A$, then we obtain a prequasihomomorphism $(\phi,\bar{\phi})$ from A to $K\otimes B$ taking for ϕ the first component of Φ and setting $\bar{\phi}(x) = F_0\phi'(x)F_0^*$, $x \in A$, where ϕ' is as above. Conjugating by $\begin{pmatrix} 1 & 0 \\ 0 & F_0 \end{pmatrix}$ one sees that (Φ,F') is homotopic to the pair

$$\begin{pmatrix} \phi & 0 \\ 0 & \overline{\phi} \end{pmatrix}, \begin{pmatrix} 0 & 1 \\ 1 & 0 \end{pmatrix}$$

As A. Connes pointed out, every (Φ,F) in $E(\ ,B)$ can be replaced by a normalized (Φ,F') such that $(F-F')\Phi(x) \in M_2(K\otimes B)$ for all $x \in A$. This can easily be seen, by lifting the image of F_0 in an appropriate quotient, to F_0' in $M_2(M(K\otimes B))$ which is "unitary" in the sense that $(F_0'^*F_0' - 1)$, $(F_0'F_0'^* - 1)$ are zero, if multiplied by $\phi(A)$, $\phi'(A)$. Applying this procedure also to elements in $E(A,B([0,1]))$ one sees that one could define $KK(A,B)$ as the set of homotopy classes of normalized elements of $E(A,B)$ with homotopies also in this category.

The isomorphism of Kasparov's $KK(A,B)$ with ours now follows from the following lemma.

5.1 <u>Lemma</u>. Let $(\phi_i,\overline{\phi}_i) : A \to E_i \triangleright J_i \subset K\otimes B$, $i = 0,1$, be prequasihomomorphisms such that the maps D_i and Q_i, obtained from $(\phi_i,\overline{\phi}_i)$ are equal for i = 0,1. Then the elements (Φ_0,F) and (Φ_1,F) in $E(A,B)$ defined by

$$\Phi_i = \begin{pmatrix} \phi_i & 0 \\ 0 & \overline{\phi}_i \end{pmatrix} \qquad i = 0,1 \qquad F = \begin{pmatrix} 0 & 1 \\ 1 & 0 \end{pmatrix}$$

are homotopic in Kasparov's sense.

<u>Proof</u>. Let $D = D_0 = D_1$. It is easily checked that the element of $E(A,B([0,\frac{\pi}{2}]))$ defined by the prequasihomomorphisms $(\psi_t,\overline{\psi}_t)$, $t \in [0,\frac{\pi}{2}]$ with

$$\psi_t = \begin{pmatrix} \phi_0 & 0 \\ 0 & \phi_1 \end{pmatrix} \quad \overline{\psi}_t = \begin{pmatrix} \phi_0 & 0 \\ 0 & \phi_1 \end{pmatrix} - W_t \begin{pmatrix} D & 0 \\ 0 & 0 \end{pmatrix} W_t^*$$

$$W_t = \begin{pmatrix} \cos t & \sin t \\ -\sin t & \cos t \end{pmatrix}$$

is a homotopy connecting $(\Phi_0,F) \oplus$ (degenerate element) to (degenerate element) $\oplus\ \Phi_1(F)$. q.e.d.

The fact that Kasparov's product coincides with ours is essentially the remark after Proposition 3.2.

Erratum:

I am grateful to D. Testard and G. Elliott, who pointed out to me that the proof of Proposition 3.1 is not conclusive as written. Here is the correct version: If $(\phi,\bar{\phi})$ already extends to E_1, let K_0, K, K' be defined as before and let

$$K_0' = \{x \in K' \mid xK \subset K_0\}$$

Then $K_0' \cap K = K_0$ and K/K_0 injects into K'/K_0'. Moreover, K'/K_0' is isomorphic to the 2×2-matrices over some C*-algebra (the image under ϕ, or $\bar{\phi}$, of E_1 in the multipliers of $\phi(J_1)/J$). Let δ be the derivation of K' that lifts the generator $\dot{\delta}$ of the one parameter group $\mathrm{Ad}\, F^t$, $t \in \mathbb{R}$, of automorphisms of K'/K_0', and let $\sigma_t = e^{\delta t}$. We have $\sigma_t(K_0') = K_0'$ and by lifting first to K'/K_0, then to K', we may also assume that $\sigma_t(K_0) = (K_0)$. Since $\sigma_t(K) \subset K + K_0'$ and $\sigma_t(K)K_0' = \sigma_t(KK_0') = \sigma_t(K_0) = K_0$ it follows that $\sigma_t(K) = K$. Now $(\mathrm{Ad}\, F^t\begin{pmatrix} 0 & 0 \\ 0 & \phi \end{pmatrix}, j\sigma_t\begin{pmatrix} 0 & 0 \\ 0 & \bar{\phi} \end{pmatrix})$, $t \in [0,1]$ is a continuous path of prequasihomomorphisms connecting $(\phi^e, \bar{\phi}^e)$ - defined using the particular $\sigma = \sigma_1$ constructed above - to $(\phi,\bar{\phi})$.

Also, one should assume that the approximate unit u_λ is quasicentral in E_1 (i.e. $u_\lambda e - eu_\lambda \to 0$ for each $e \in E_1$, cf. [5,3.12.14]). Then this path consists entirely of extendible prequasihomomorphisms.

References.

1. A. Connes and G. Skandalis, The longitudinal index theorem for foliations, preprint.
2. J. Cuntz, On the homotopy groups of the space of endomorphisms of a C*-algebra, Proc. of the OAGR conference at Neptun, Romania, to appear.
3. J. Cuntz, K-theory and C*-algebras, Proc. of the K-theory conference at Bielefeld, 1982, to appear.
4. G.G. Kasparov, The operator K-functor and extensions of C*-algebras, Izv. Akad. Nauk SSSR, Ser. Mat. 44(1980), 571-636.
5. G.K. Pedersen, C*-algebras and their automorphism groups, Academic Press, London-New York-San Francisco, 1979.
6. G. Skandalis, Some remarks on Kasparov theory, preprint.

Global Equilibria and Steady States of Discrete Networks, according to Classical Thermodynamics

A Phenomenological Approach to and Evaluation of Prigogine's Principle of Minimum Entropy Production

Andreas Dress
Fakultät für Mathematik
Universität Bielefeld
4800 Bielefeld 1
Federal Rebublic of Germany

Abstract: Local and global equilibrium states of composite thermodynamic systems are discussed from a phenomenological "Gibbsian" point of view with special emphasis on the principle of minimum entropy production, starting from a system of linear differential equations modelling the evolution of such systems from local to global equilibrium or steady states caused by diffusion.

§ 1 Gibb's Description of Simple Thermodynamic Systems

According to Gibbs (cf. [1] - [4]) a simple thermodynamic system Σ with $n+1$ degrees of freedom can be described by its "fundamental equation", i. e. by a function $s = s_\Sigma = s_\Sigma(x_o,..,x_n)$ of n+1 variables, positively homogeneous of degree 1 and piecewise twice differentiable, defined on an open cone in $\mathbb{R}^{n+1}$. As usual, the first variable x_o stands for the internal energy of Σ, x_1 for its volume and s for its entropy.

From the homogeneity of s it follows that the hessian matrix $\left(\frac{\partial^2 s}{\partial x_1\ \partial x_j}\right)$, wherever it is defined, cannot have rank $n+1$, since

$$(1) \qquad \sum_{j=0}^{n} x_j \frac{\partial^2 s}{\partial x_i\ \partial x_j} = 0$$

holds for all $i = 0,1,\dots,n$. A state of equilibrium Φ of Σ is any point $(x_o,x_1,\dots,x_n) \in \mathbb{R}^{n+1}$ in the domain of definition of s such that s is twice differentiable and the hessian $\left(\frac{\partial^2 s}{\partial x_i\ \partial x_j}\right)$ is negative semi-definite of rank n at $(x_o,x_1,\dots,x_n)$ - i. e. s is strictly locally convex at Φ. We denote the set or rather the differentiable manifold of equilibrium states of Σ by D_Σ . For any equilibrium state $\Phi = (x_o, x_1, \dots, x_n) \in D_\Sigma$ we have the tangent space T_Φ at Φ , i. e. the space of infinitesimal quasi-static processes starting (or terminating) in Φ , and its dual, the cotangent space $\hat{T}_\Phi$ at Φ , spanned by the differentials df of the differentiable functions $f : D_\Sigma \to \mathbb{R}$, i. e. the "variables of state" of Σ .

Any such $f : D_\Sigma \to \mathbb{R}$ is said to be an extensive or an intensive variable of state, if f is homogeneous of degree 1 or 0, respectively. In particular, the entropy

$$S : D_\Sigma \to \mathbb{R} : \Phi = (x_o, \dots, x_n) \longmapsto S(\Phi) =: s(x_o, \dots, x_n)$$

and the coordinate functions

$$X_i : D_\Sigma \to \mathbb{R} : \Phi = (x_o, \dots, x_n) \longmapsto X_i(\Phi) =: x_i$$

are extensive variables, whereas the so-called "generalized pressures" given by the partial derivatives

$$p_i : D_\Sigma \to \mathbb{R} : \Phi = (x_o, \dots, x_n) \longmapsto \frac{\partial s}{\partial x_i}(x_o, \dots, x_n)$$

are intensive variables.

In the cotangent space $\hat{T}_\Phi$ of an equilibrium state $\Phi \in D_\Sigma$ the relation (1) can be expressed in the form

$$\sum_{i=0}^{n} X_i(\Phi) dp_i = 0 , \tag{2}$$

the Gibbs-Duhem relation.

Moreover, from our assumption that the rank of the matrix $\left(\frac{\partial^2 s}{\partial x_i \, \partial x_j}\right)$ is precisely n for any $\Phi \in D_\Sigma$, it follows that the subspace of $\hat{T}_\Phi$ which is generated by the differentials of all intensive variables is of dimension n and generated by $dp_o, dp_1, \dots$ $\dots, dp_n$. In particular, if $\sum_{i=0}^{n} c_i dp_i = 0$, then there exists $\lambda \in \mathbb{R}$ with $c_i = \lambda x_i$ for all $i = 0, 1, \dots, n$.

Also, it follows from the homogeneity of s that the fundamental equation can be written in the form

$$S(\Phi) = \sum_{i=0}^{n} X_i(\Phi) \cdot p_i(\Phi) \quad . \tag{3}$$

§ 2 Thermodynamics of Composite Systems, Local and Global Equilibria

In the following we consider a composite thermodynamic system Σ , consisting of N simple subsystems $\Sigma^\alpha (\alpha = 1, 2, \dots, N)$. For simplicity we assume all the systems to have

the same degree $n+1$ of freedom and we assume their entropy functions $s^\alpha =: s_{\Sigma^\alpha}$ to depend on the same variables $x_o, x_1, \dots, x_n$, but we will allow the s^α to differ from each other and thus have individual manifolds of equilibrium states $D^\alpha =: D_{\Sigma^\alpha}$ $(\alpha = 1,2,\dots,N)$. We define a local equilibrium state Φ of Σ to consist of a family $(\Phi^\alpha)_{\alpha=1,2,\dots,N}$ of equilibrium states of the various subsystems, i. e. we identify the manifold D_Σ^{loc} of local equilibrium states of Σ with the product $\prod_{\alpha=1}^N D^\alpha$.
Since the multiplicative group $\mathbb{R}_+^\times$ of positive real numbers acts on each factor D^α (by $\lambda \cdot \Phi^\alpha = \lambda(x_o^\alpha, x_1^\alpha, \dots, x_n^\alpha) =: (\lambda x_o^\alpha, \lambda x_1^\alpha, \dots, \lambda x_n^\alpha)$, $\lambda > 0$), the direct product $\prod_{\alpha=1}^N \mathbb{R}_+^\times$ of N copies of $\mathbb{R}_+^\alpha$ acts on D_Σ^{loc} by $\Lambda \cdot \Phi = (\lambda_1, \dots, \lambda_N) \cdot (\Phi^1, \dots, \Phi^N) =:$
$=: (\lambda_1 \Phi^1, \dots, \lambda_N \Phi^N)$.

Again, we define a function $f : D_\Sigma^{loc} \to \mathbb{R}$ to be intensive, if $f(\Lambda \cdot \Phi) = f(\Phi)$ for all $\Lambda \in \prod_{\alpha=1}^N \mathbb{R}_+^\times$ and all $\Phi \in D_\Sigma^{loc}$, and we define f to be extensive if there exist functions $f^\alpha : D_\Sigma^{loc} \to \mathbb{R}$ $(\alpha = 1, \dots, N)$ with $f(\Lambda \cdot \Phi) = \sum_{\alpha=1}^N \lambda_\alpha f^\alpha(\Phi)$ for all $\Lambda = (\lambda_1, \dots, \lambda_N)$ and all Φ.

Examples of intensive variables are again the generalized pressures

$$p_i^\alpha : D_\Sigma^{loc} \to \mathbb{R} : \Phi = (\Phi^1, \dots, \Phi^N) \mapsto p_i(\Phi^\alpha) ,$$

examples of extensive variables are the coordinate functions

$$X_i^\alpha : D_\Sigma^{loc} \to \mathbb{R} : \Phi = (\Phi^1, \dots, \Phi^n) \mapsto X_i(\Phi^\alpha)$$

as well as the entropies of the various subsystems

$$S^\alpha : D_\Sigma^{loc} \to \mathbb{R} : \Phi = (\Phi^1, \dots, \Phi^n) \mapsto S(\Phi^\alpha)$$

and their sum, the total entropy,

$$S = S_\Sigma : D_\Sigma^{loc} \to \mathbb{R} : \Phi \mapsto \sum_{\alpha=1}^N S^\alpha(\Phi) .$$

The problem I want to discuss is how to describe mathematically the set (manifold) of global equilibrium or - at least - steady states of Σ and the way these states are attained. Of course, this depends strongly on the interaction we suppose to take place between the various subsystems. In the following we restrict our attention to processes of diffusion, only. Thus we assume that for any two neighbouring subsystems Σ^α and Σ^β the exchange of the quantity measured by X_i $(i=0,\dots,n)$ is proportional to the difference of the generalized pressures p_i^α and p_i^β in Σ^α and Σ^β which are associated to X_i. Thus we assume the quantity X_i to vary according to the differential equation

(4) $$\dot{X}_i^\alpha = \sum_{\beta=1}^{N} c_i^{\alpha\beta}(p_i^\alpha - p_i^\beta)$$

with some non-negative proportionality factors $c_i^{\alpha\beta}$ $(i=0,1,\ldots,n\ ;\ \alpha,\beta = 1,2,\ldots,N)$. We put $c_i^{\alpha\alpha} = 0$ for all i and α and we have $c_i^{\alpha\beta} = 0$ if either Σ^α and Σ^β are spatially disconnected so that no direct interaction by diffusion can take place or if Σ^α and Σ^β are neighbouring subsystems, but the exchange of the quantity measured by X_i is prohibited by an appropriate wall between Σ^α and Σ^β. Otherwise we have $c_i^{\alpha\beta} > 0$. Since whatever flows from Σ^β to Σ^α and has to be added there has to be deducted from Σ^β at the same time (and vice versa), we have

(5) $$c_i^{\alpha\beta} = c_i^{\beta\alpha} \quad .$$

At this stage we do not assume the $c_i^{\alpha\beta}$ to be constant.

We define two subsystems Σ^α and Σ^β to be i-connected, if $\alpha = \beta$ or if $c_i^{\alpha\beta} \neq 0$ or if - more generally - there exists a string of indices $\alpha = \alpha_0, \alpha_1, \ldots, \alpha_k = \beta$ with $c_i^{\alpha_{\kappa-1}\alpha_\kappa} \neq 0$ for all $\kappa = 1,\ldots,k$. We define a subset $A \subseteq \{1,2,\ldots,N\}$ of indices (subsystems) to be an i-connected component, if A is non-empty and consists of all α which are i-connected to some (any) given $\beta \in A$. We define two local equilibrium states Φ and Φ' to be of the same overall composition if $\sum_{\alpha\in A} X_i^\alpha(\Phi) = \sum_{\alpha\in A} X_i^\alpha(\Phi')$ for all i-connected components $A \subseteq \{1,\ldots,N\}$ and all $i \in \{0,\ldots,n\}$. Using this terminology we can state the following (of course very well-known) result

THEOREM 1: For any $\Phi \in D_\Sigma^{loc}$ the following statements are equivalent:

(i) Φ is a steady state, i. e.

$$\dot{X}_i^\alpha(\Phi) = \sum_\beta c_i^{\alpha\beta}(\Phi)(p_i^\alpha(\Phi) - p_i^\beta(\Phi)) = 0 \text{ for all } i \text{ and } \alpha .$$

(ii) For any $(i;\alpha,\beta)$ with $c_i^{\alpha\beta} \neq 0$ we have $p_i^\alpha(\Phi) = p_i^\beta(\Phi)$.

(iii) Φ is a steady state for S_Σ, i. e. $\dot{S}_\Sigma(\Phi) = 0$.

(iv) $S(\Phi) \geq S(\Phi')$ for all $\Phi' \in D_\Sigma^{loc}$ of the same overall composition, for which $\lambda\Phi + (1-\lambda)\Phi' \in D_\Sigma^{loc}$ for all $\lambda \in [0,1]$ (with $\lambda\Phi$ denoting the state defined by $X_i^\alpha(\lambda\Phi) = \lambda X_i^\alpha(\Phi)$) in particular S has a local maximum at Φ relative to the set of $\Phi' \in D_\Sigma^{loc}$ with the same overall composition.

Proof: "(ii) $\Rightarrow$ (i)" is trivial and "(i) $\Rightarrow$ (iii)" holds since

$$\dot{S}(\Phi) = \sum_\alpha \dot{S}^\alpha(\Phi) = \sum_\alpha \sum_i \frac{\partial s^\alpha}{\partial x_i} \cdot \dot{X}_i^\alpha = \sum_\alpha \sum_i p_i^\alpha(\Phi)\cdot\dot{X}_i^\alpha(\Phi) .$$

Using (4) we get moreover

$$\dot{S}(\Phi) = \sum_\alpha \sum_i p_i^\alpha(\Phi) \sum_\beta c_i^{\alpha\beta}(p_i^\alpha(\Phi) - p_i^\beta(\Phi))$$

and thus - using $c_i^{\alpha\beta} = c_i^{\beta\alpha}$ -

(6) $$\dot{S}(\Phi) = \frac{1}{2} \sum_i \sum_{\alpha,\beta} c_i^{\alpha,\beta}(p_i^\alpha(\Phi) - p_i^\beta(\Phi))^2 .$$

This shows $\dot{S}(\Phi) \geq 0$ for all $\Phi \in D_\Sigma^{loc}$ and $\dot{S}(\Phi) = 0$ if and only if (ii) holds.

To prove "(iv) ⇒ (iii)" it is enough to remark that for any solution $\Phi(t)$ of (4) all the states $\Phi(t)$ have the same overall composition, since for any i-connected component A we have

$$\sum_{\alpha\in A} \dot{X}_i^\alpha = \sum_{\alpha\in A} \sum_{\beta=1}^{N} c_i^{\alpha\beta}(p_i^\alpha - p_i^\beta) = \sum_{\alpha,\beta\in A} c_i^{\alpha\beta}(p_i^\alpha - p_i^\beta) = \sum_{\substack{\alpha,\beta\in A\\ \alpha<\beta}} (c_i^{\alpha\beta} - c_i^{\beta\alpha})(p_i^\alpha - p_i^\beta) = 0$$

the second equality being true since $c_i^{\alpha\beta} = 0$ for $\alpha \in A$ and $\beta \notin A$.

To prove "(ii) ⇒ (iv)" we consider

$$f(\lambda) =: S(\lambda\Phi + (1-\lambda)\Phi') \quad (\lambda \in [0,1]) .$$

We have to show that $f'(\lambda) \geq 0$ which follows from the fact that

$$f'(\lambda) = \sum_\alpha \sum_i p_i^\alpha(\lambda\Phi + (1-\lambda)\Phi')(X_i^\alpha(\Phi) - X_i^\alpha(\Phi'))$$

is itself monotonously not increasing since

$$f''(\lambda) = \sum_\alpha \sum_{i,j} \frac{\partial^2 s}{\partial x_i \partial x_j}(X_i^\alpha(\Phi) - X_i^\alpha(\Phi'))(X_j^\alpha(\Phi) - X_j^\alpha(\Phi')) \leq 0$$

by the convexity of the s^α together with

$$f'(1) = \sum_\alpha \sum_i p_i^\alpha(\Phi)(X_i^\alpha(\Phi) - X_i^\alpha(\Phi')) = \sum_i \sum_p p \sum_{\substack{\alpha\\ p_i^\alpha(\Phi)=p}} (X_i^\alpha(\Phi) - X_i^\alpha(\Phi')) = 0 ,$$

the last equality being true since by (ii) the set $A_{i,p} = \{\alpha \mid p_i^\alpha(\Phi) = p\}$ consists of full i-connected components, so we have $\sum_{\alpha\in A_{i,p}} X_i^\alpha(\Phi) = \sum_{\alpha\in A_{i,p}} X_i^\alpha(\Phi')$ using the fact that Φ and Φ' have the same overall composition.

<u>Remark 1</u>: It follows from this proof that for any Φ , satisfying (ii), we have $S(\Phi) > S(\Phi')$ for any Φ' close enough to Φ and of the same overall composition once we know that

$$f''(1) = \sum_\alpha \sum_{i,j} \frac{\partial^2 s}{\partial x_i \partial x_j}(\Phi)(X_i^\alpha(\Phi) - X_i^\alpha(\Phi'))(X_j^\alpha(\Phi) - X_j^\alpha(\Phi')) < 0 .$$

But it follows from the discussion at the end of § 1 that $f''(1) = 0$ holds if and only if $X_i^\alpha(\Phi) - X_i^\alpha(\Phi') = c_\alpha X_i^\alpha(\Phi)$ for an appropriate family of real numbers c_α ($\alpha = 1,\dots,N$) which are independent of i. Thus $f''(1) = 0$ if and only if $X_i^\alpha(\Phi') = \lambda_\alpha X_i^\alpha(\Phi)$, i. e. $\Phi^{\alpha'} = \lambda_\alpha \Phi^\alpha$ and $\Phi' = \Lambda\cdot\Phi$ for some family $\Lambda = (\lambda_1,\dots,\lambda_N) \in \mathbb{R}^N$.

If we introduce enough parameter-specific impermeable walls between our various subsystems to distinguish them from each other (e. g. by defining each Σ^α to be the subsystem, contained in a well-defined area of our system Σ so that volume cannot be exchanged by the Σ^α by their very definition, or - in case we deal with rigid bodies - to contain a well-defined and well-bounded portion of the matter, constituting Σ, so that matter cannot be exchanged by the Σ^α by their very definition), it follows that Φ and $\Phi' = \Lambda\cdot\Phi$ cannot be of the same overall constitution unless $\Phi' = \Phi$.
(For a more detailed discussion of the problems related to this question see the appendix.) Thus we can conclude that in general, i. e. for "well-defined" subsystems the entropy function is indeed a Ljapunov-function for diffusion processes and that among all local equilibrium states of the same overall constitution the global equilibrium state is characterized as the state with maximal entropy, - at least as long as the set of local equilibrium states is a convex subset of $\mathbb{R}^{n\cdot N}$ with respect to the canonical embedding

$$D_\Sigma^{loc} \to \mathbb{R}^{n\cdot N} : \Phi \longrightarrow (X_i^\alpha(\Phi))_{i=1,\dots,n;\ \alpha=1,\dots,N}$$

(or any other embedding in some $\mathbb{R}^m$, attained by extensive variables of state).

Remark 2: Note that (6) does not imply that the entropy production $\dot{S}^\alpha(\Phi)$ of the various subsystems is non-negative for all α ; in the opposite, it is rather easy to construct composite systems Σ and to find local equilibrium states $\Phi \in D_\Sigma^{loc}$ such that $\dot{S}^\alpha(\Phi)$ is negative.

§ 3 The Principle of Minimum Entropy Production

Let us suppose that some of the subsystems Σ^α are not proper simple thermodynamic systems but heat etc. bathes, which means that though they interact with other subsystems or even with each other and thus exchange some of the quantities measured by the X_i, varying their extensive variables, they all contain these quantities to so high a degree, that these changes of the X_i do not affect the values of their intensive variables. Thus - according to (3) - we model the fundamental equation of a heat bath, whose intesive variables $p_o,p_1,\dots,p_n$ are fixed at certain constant values $q_o,q_1,\dots,q_n$, by its linear approximation

(7) $$s(x_o,\ldots,x_n) = \sum_{i=0}^{n} q_i x_i + c \quad .$$

Though the additive constant c in this equation remains undetermined, this does not matter as long as we are interested only in relating the changes of the X_i to changes of S, i. e. we may even put $c = 0$ or we may write (7) in its differential form

(7') $$dS = \sum_{i=0}^{n} q_i dX_i \quad .$$

Now assume that only the subsystems Σ^α for $\alpha = 1,2,\ldots,M$ are proper simple thermodynamic systems for some $M \leq N$ whereas the subsystems Σ^α for $\alpha = M+1,\ldots,N$ are heat etc. bathes, whose intensive variables are fixed at certain values $q_o^\alpha, q_1^\alpha, \ldots, q_n^\alpha$. It follows from (6) that the entropy production $\dot{S}(\Phi)$ will be positive for all local equilibrium states $\Phi \in D_\Sigma^{loc}$ once there exist at least two heat bathes Σ^α and Σ^β in the same i-connected component for some i with $q_i^\alpha \neq q_i^\beta$. Thus we cannot expect to characterize steady states by maximal entropy.

On the other hand, it seems reasonable to define a steady state Σ not by requiring $\dot{X}_i^\alpha = 0$ for all i and α, but only for $\alpha = 1,2,\ldots,M$, leaving the various heat bathes aside.

Let us now consider equation (6) as an equation which expresses the total entropy production $\dot{S}(\Phi)$ as a quadratic form in the $p_i^\alpha(\Phi)$ with coefficients determined by the $c_i^{\alpha\beta}$ and the q_i^α. We define Σ to be a Prigogine system - at least in some area of D_Σ^{loc} - if all the $c_i^{\alpha\beta}$ are constant - at least in this area.

In this case it is easy to see that the always positive quadratic form

(8) $$\begin{aligned} Q(p_i^\alpha) = {} & \frac{1}{2} \sum_{i=0}^{n} \sum_{\alpha,\beta=1}^{M} c_i^{\alpha\beta}(p_i^\alpha - p_i^\beta)^2 \\ & + \sum_{i=0}^{n} \sum_{\alpha=1}^{M} \sum_{\beta=M+1}^{N} c_i^{\alpha\beta}(p_i^\alpha - q_i^\beta)^2 \\ & + \frac{1}{2} \sum_{i=0}^{n} \sum_{\alpha,\beta=M+1}^{N} c_i^{\alpha\beta}(q_i^\alpha - q_i^\beta)^2 \end{aligned}$$

will attain its minimum for those p_i^α for which the partial derivatives

(9) $$\frac{\partial Q}{p_j^\beta}(p_i^\alpha) = 2 \sum_{\alpha=1}^{M} c_j^{\alpha\beta}(p_j^\beta - p_j^\alpha) + 2 \sum_{\alpha=M+1}^{N} c_j^{\alpha\beta}(p_j^\beta - q_j^\alpha)$$

equal 0.

But a comparison with (4) shows

$$(10) \qquad \frac{\partial Q}{\partial p_j^\beta}(p_i^\alpha(\Phi)) = 2\,\dot{X}_j^\beta(\Phi)\ .$$

Thus we see that $\Phi \in D_\Sigma^{loc}$ represents a steady state if and only if the entropy production $\dot{S}(\Phi)$ at Φ is at its minimal value - compared with its value at all imaginable combinations of the generalized pressure of the subsystems $\Sigma^1, \Sigma^2, \ldots, \Sigma^M$.

To show that, vice versa, a state $\Phi_o \in D_\Sigma^{loc}$ is a steady state if $\dot{S}(\Phi_o)$ is minimal compared with its values at close by local equilibrium states Φ' which are of the same overall composition as Φ, we consider the solution $\Phi(t)$ of (4) with $\Phi(0) = \Phi_o$ and compute

$$(11) \qquad \ddot{S}(\Phi) = \sum_{i=0}^{n} \sum_{\alpha,\beta=1}^{M} c_i^{\alpha\beta}(p_i^\alpha - p_i^\beta) \left(\sum_j \frac{\partial^2 s^\alpha}{\partial x_i\, \partial x_j} \dot{X}_j^\alpha - \sum_j \frac{\partial^2 s^\alpha}{\partial x_i\, \partial x_j} \dot{X}_j^\beta \right)$$

$$+ 2 \sum_{i=0}^{n} \sum_{\alpha=1}^{M} \sum_{\beta=M+1}^{N} c_i^{\alpha\beta}(p_i^\alpha - q_i^\beta) \sum_j \frac{\partial^2 s^\alpha}{\partial x_i\, \partial x_j} \dot{X}_j^\alpha$$

$$= 2 \sum_{\alpha=1}^{M} \sum_{i,j} \frac{\partial^2 s_\alpha}{\partial x_i\, \partial x_j} \dot{X}_i^\alpha \dot{X}_j^\alpha\ .$$

It follows that $\ddot{S}(\Phi) \leq 0$ and $\ddot{S}(\Phi) = 0$ if and only if $\dot{X}_i^\alpha = \lambda_\alpha \cdot X_i^\alpha$ for some appropriate family of parameters $\lambda_1, \ldots, \lambda_M$. To show that this implies $\dot{X}_i^\alpha = 0$ we introduce once more the hypothesis that for any subsystem Σ^α there exists an index i such that $c_i^{\alpha\beta} = 0$ for all $\beta = 1,2,\ldots,N$, i. e. there exists a quantity measured by some X_i which Σ^α does not exchange with any other subsystem Σ^β. This implies $\dot{X}_i^\alpha = 0$ for this particular i and therefore $\lambda_\alpha X_i^\alpha = 0$, which in turn implies $\lambda_\alpha = 0$ (assuming $X_i^\alpha(\Phi) \neq 0$) and therefore $\dot{X}_j^\alpha = \lambda_\alpha X_j^\alpha = 0$ for all $j = 0,1,\ldots,n$.

Thus we have proved

<u>THEOREM 2</u>: If Σ is a Prigogine system, consisting of M simple subsystems $\Sigma^1, \ldots, \Sigma^M$ and some heat etc. bathes $\Sigma^{M+1}, \ldots, \Sigma^N$, and if we assume that for any of the simple subsystems Σ^α $(\alpha = 1,\ldots,M)$ there exists an index $i = i(\alpha)$ with $c_i^{\alpha\beta} = 0$ for all $\beta = 1,2,\ldots,N$, then a local equilibrium state $\Phi \in D_\Sigma^{loc}$ with $X_{i(\alpha)}^\alpha(\Phi) \neq 0$ for $\alpha = 1,\ldots,M$ is a steady state if and only if the entropy production $\dot{S}(\Phi)$ has a (local) minimum at Φ.

If Φ is not a steady state, then $\ddot{S}(\Phi)$ will be negative, so $\dot{S}(\Phi)$ will be a monotonously decreasing function and $\Phi(t)$ will thus approach a steady state.

Remark 1: The assumption that $c_i^{\alpha\beta}$ is constant is rather essential. If $i = 0$ we have $p_o^\alpha = \frac{1}{T^\alpha}$ with T^α denoting the temperature at Σ^α.

Thus our assumption amounts to the following assertion for $i = 0$: If two neighbouring thermodynamic systems of different temperature T_1 and T_2 are allowed to exchange heat, then the rate at which internal energy in the form of heat will be transferred from the first to the second is proportional to $\frac{1}{T_1} - \frac{1}{T_2}$, and not to, say, $T_2 - T_1 = T_1 \cdot T_2 \left(\frac{1}{T_1} - \frac{1}{T_2}\right)$, which, at a first glance, might be as plausible an *Ansatz* as the formula we have used, but would lead to the non-constant proportionality factor $c_o^{12} = T_1 \cdot T_2 \cdot c$, which in turn would not allow to derive the principle of minimum entropy production. Similar consideration hold for the other generalized pressures.

Thus it seems worthwhile to check this assumption experimentally.

Remark 2: Similarly the assumption that the subsystems $\Sigma^1,\ldots,\Sigma^M$ interact with the subsystems $\Sigma^{M+1},\ldots,\Sigma^N$ by diffusion, only, is rather essential. Once we would suppose, modelling a chemical tank reactor, that the systems $\Sigma^{M+1},\ldots,\Sigma^N$ interact with the systems $\Sigma^1,\ldots,\Sigma^M$ by pumping certain quantities at a constant or otherwise controlled rate into or out of the subsystems we cannot expect the principle of minimum energy production to hold any longer as a characterization of steady states.

Remark 3: It should not be too difficult to go to the limit of discrete diffusion networks by subdividing a given system into finer and finer subsystems and thus to use this approach to study continuous diffusion systems, too.

References

[1] J. W. Gibbs, Graphical Methods in the Termodynamics of Fluids. Transaction of the Connecticut Academy II (1873) 3/9-342 Coll.

[2] __________, A Method of Geometrical Representation of the Thermodynamic Properties of Substances by Means of Surfaces. Transactions of the Connecticut Academy II (1873) 382-404 Coll.

[3] H. B. Callen, Thermodynamics, an introduction to the physical theories of equilibrium thermostatics and irreversible thermodynamics. John Wiley, New York, 1960.

[4] A. S. Wightman, Convexity and the notion of equilibrium state in thermodynamics and statistical mechanics. Introduction to: R. B. Israel: Convexity in the theory of lattice gases, Princeton University Press, Princeton, N. J., 1979.

Appendix
The mathematical formalism of equilibrium thermodynamics of composite systems.

It is considered to be folklore among people studying classical thermodynamics that there is no essential and at least no formal difference between simple and composite closed thermodynamic systems. It is my purpose to reconsider this judgement from a mathematical point of view - starting from our definition of an equilibrium state of a simple thermodynamic system Σ with the fundamental equation $s = s_\Sigma(s_o,\dots,x_n)$ as a state $\Phi = (x_o,x_1,\dots,x_n) \in \mathbb{R}^{n+1}$ at which s is twice differentiable and negative semidefinite of maximal rank n - so there is only one relation between the differentials $dp_o,dp_1,\dots,dp_n \in \hat{T}_\Phi$ at Φ, the Gibbs-Duhem relation.

Let us now consider a composite system $\Sigma = (\Sigma^1,\Sigma^2,\dots,\Sigma^N)$ - using the notation developed in § 2. We suppose all the Σ^α to be simple thermodynamic systems, whose interaction is given by a family of non-negative parameters $c_i^{\alpha\beta}$ $(i = 0,\dots,n\ ;\ \alpha,\beta = 1,\dots,N)$ which we do not assume to be constant, but to be either 0 or positive all over. According to Theorem 1 we define a local equilibrium state $\Phi = (\Phi^1,\dots,\Phi^N) \in D_\Sigma^{loc} = \prod_{\alpha=1}^N D_{\Sigma^\alpha}$ to be a global equilibrium state, if $p_i^\alpha(\Phi) = p_i^\beta(\Phi)$ for all $(i;\alpha,\beta)$ with $c_i^{\alpha\beta} \neq 0$ or - equivalently - if $p_i^\alpha(\Phi)$ has the same value for all α in any given i-connected component $A \subseteq \{1,2,\dots,N\}$. We denote the set of global equilibrium states of Σ by D_Σ. Generally, a global equilibrium state $\Phi \in D_\Sigma$ should be completely determined by its overall composition, i. e. by the values $X_i^A(\Phi) =: \sum_{\alpha\in A} X_i^\alpha(\Phi)$ with i running through $0,1,\dots,n$ and $A \subseteq \{1,2,\dots,N\}$ for each i through all i-connected components of Σ. But, of course, this can be expected only if the subdivision of Σ into subsystems is not completely arbitrary.

Consider for instance the classical simple thermodynamic system, a fluid in a closed cylinder, whose equilibrium states are determined by the variables U (internal energy), V (volume) and N (matter). If we subdivide in a *Gedankenexperiment* this Σ arbitrarily into two subsystems Σ^1 and Σ^2 which are allowed to exchange everything, energy, volume and matter, the values of $U = U^1 + U^2$, $V = V^1 + V^2$ and $N = N^1 + N^2$ will determine the associated generalized pressures for both subsystems, but they will determine (U^1,V^1,N^1) and (U^2,V^2,N^2) up to a scalar factor only, i. e. all the global equilibrium states $(\lambda U^1,\lambda V^1,\lambda N^1\ ;\ (1-\lambda)U^2,(1-\lambda)V^2,(1-\lambda)N^2)$ define essentially the same equilibrium state. Of course, if (- in the same *Gedankenexperiment* -) we associate to Σ^1 and Σ^2 a fixed volume which cannot be exchanged any more, e. g. by introducing a firmly rigid piston somewhere with a small hole in its middle, or if we introduce an impermeable and rigid, but movable piston between our two systems, this way blocking the exchange of matter, we are out of trouble: If U,V_1,V_2 and N are given, we will get $U_i = \frac{V_i}{V_1 + V_2} \cdot U$, $N_i = \frac{V_i}{V_1 + V_2} \cdot N$ $(i=1,2)$; if U,V,N_1,N_2 are given, we get similarly $U_i = \frac{N_i}{N_1 + N_2} \cdot U$, $V_i = \frac{N_i}{N_1 + N_2} \cdot V$ $(i=1,2)$.

In general, we define $\Phi \in D_\Sigma$ to possess free play, if there exists a family of real numbers $(\lambda_\alpha)_{\alpha=1,2,\dots,N} \neq (0,\dots,0)$ with $\sum_{\alpha\in A} \lambda_\alpha X_i^\alpha(\Phi) = 0$ for all $i = 0,1,\dots,n$ and all i-connected components $A \subseteq \{1,\dots,N\}$. The set (vector space) of all those parameters $(\lambda_\alpha)_{\alpha=1,2,\dots,N}$ will be denoted by Pl_Φ – the "play of Φ" . Φ is defined to be without play, if $\dim Pl_\Phi = 0$. It is obvious that Φ has no play if for any $\alpha = 1,2,\dots,N$ there exists an $i \in \{0,1,\dots,n\}$ such that $X_i^\alpha(\Phi) \neq 0$ and $c_i^{\alpha\beta} = 0$ for all $\beta = 1,2,\dots,N$. If we denote the set of i-connected components $A \subseteq \{1,2,\dots,N\}$ by $\mathcal{A}_i$, we can describe Pl_Φ as the kernel of the linear map

$$r_\Phi : \mathbb{R}^N \to \prod_{i=0}^{n} \mathbb{R}^{\mathcal{A}_i} : (\lambda_\alpha)_{\alpha=1,2,\dots,N} \mapsto \Big(\sum_{\alpha\in A} \lambda_\alpha X_i^\alpha(\Phi)\Big)_{\substack{i=0,1,\dots,n\\ A\in\mathcal{A}_i}}$$

For $(\lambda_\alpha)_{\alpha=1,2,\dots,N} \in Pl_\Phi$ the state Φ and the state $(1+\varepsilon\lambda_\alpha)_{\alpha=1,2,\dots,N}\cdot\Phi$ will have the same overall composition for all ε , for which $1+\varepsilon\lambda_\alpha$ is positive for all α , and the same intensive variables. Thus they denote the same equilibrium state of Φ . Vice versa, it follows from § 1, that for any two global equilibrium states Φ and Φ' with the same overall composition there exist $\varepsilon \in \mathbb{R}$ and $(\lambda_\alpha) \in Pl_\Phi$ with $\Phi' = (1+\varepsilon\lambda_\alpha)\Phi$. In particular, both systems have the same total entropy.

It is easy to see that the set of global equilibrium states without play forms an open (perhaps empty) subset of D_Σ . We want to show that this subset D_Σ^o of D_Σ , considered as a subset of $\mathbb{R}^{(1+n)\cdot N}$ via

$$D_\Sigma^o \to \mathbb{R}^{(n+1)N} : \Phi \mapsto (X_i^\alpha(\Phi))_{\substack{i=0,1,\dots,n\\ \alpha=1,\dots,N}}$$

is a differentiable submanifold of dimension $\sum_{i=0}^{n} \# \mathcal{A}_i$, which is mapped onto an open subset of $\prod_{i=0}^{n} \mathbb{R}^{\mathcal{A}_i}$ by the local diffeomorphism

$$\sigma : D_\Sigma^o \quad \prod_{i=0}^{n} \mathbb{R}^{i} : \Phi \mapsto \Big(\sum_{\sigma\in A} X_i^\alpha(\Phi)\Big)_{\substack{i=0,\dots,n\\ A\in\mathcal{A}_i}} .$$

This follows immediately from the following two statements:

<u>Theorem 3</u>: For each $\Phi \in D$ we have an exact sewuence

$$0 \longrightarrow Pl_\Phi \to \mathbb{R}^N \xrightarrow{r_\Phi} \prod_{i=0}^{n} \mathbb{R}^{\mathcal{A}_i} \to \hat{T}_\Phi(D_\Sigma) \to \hat{T}_\Phi(e^{\mathbb{R}^N}\cdot\Phi) \to 0$$

with

$$\hat{T}_\Phi(D_\Sigma) = T_\Phi(D_\Sigma^{loc})/\langle dp_i^\alpha - dp_i^\beta \mid c_i^{\alpha\beta} \neq 0 \rangle \quad \text{and} \quad \hat{T}_\Phi(e^{\mathbb{R}^N}\cdot\Phi)$$

denoting the cotangent space at Φ with respect to the submanifold

$$e^{\mathbb{R}^N}\cdot\Phi = \{(e^{\lambda_\alpha})_{\alpha=1,\ldots,N}\cdot\Phi \mid (\lambda_\alpha)_{\alpha=1,\ldots,N} \in \mathbb{R}^N\} ,$$

r_Φ being defined as in the definition of PL_Φ , $\hat{T}_\Phi(D_\Sigma) \to T_\Phi(e^{\mathbb{R}^N}\cdot\Phi)$ by the embedding $e^{\mathbb{R}^N}\cdot\Phi \longrightarrow D_\Sigma \subseteq D_\Sigma^{loc}$ and $\prod_{i=0}^{n} \mathbb{R}^{A_i} \to T_\Phi(D_\Sigma)$ by

$$(x_i^A)_{i=0,\ldots,n}^{A\in A_i} \longrightarrow \sum_{i=0}^{n} \sum_{A\in A_i} x_i^A \, d\, p_i^A ,$$

$d\, p_i^A$ denoting the image of $d\, p_i^\alpha$ in $\hat{T}_\Phi(D_\Sigma)$ for some/any $\alpha \in A$.

Theorem 4: The sequence

$$0 \to T_{\sigma(\Phi)}(\prod_{i=0}^{n} \mathbb{R}^{A_i}) \to T_\Phi(D_\Sigma) \to T_\Phi(e^{Pl_\Phi}\cdot\Phi) \to 0$$

is always exact.

Proof of Theorem 3: $0 \to Pl_\Phi \longrightarrow \mathbb{R}^N \to \prod_{i=0}^{n} \mathbb{R}^{A_i}$ is exact by the definition of PL_Φ . The composition of the maps

$$\mathbb{R}^N \xrightarrow{r_\Phi} \prod_{i=0}^{n} \mathbb{R}^{A_i} \to T_\Phi(D_\Sigma)$$

is the zero map because of the Gibbs-Duhem relation: A parameter family $(\lambda_1,\ldots,\lambda_N)$ is mapped onto

$$\sum_{i=0}^{n} \sum_{A\in A_i} \sum_{\alpha\in A} \lambda_\alpha X_i(\Phi) dp_i^A = \sum_{\alpha=1}^{N} \lambda_\alpha \sum_{i=0}^{n} X_i^\alpha(\Phi) dp_i^\alpha = 0$$

Assume, vice versa, that $\sum_{i=0}^{n} \sum_{A\in A_i} X_i^A \, d\, p_i^A = 0$ in $\hat{T}_\Phi(D_\Sigma)$. Choose a representative $\alpha_A \in A$ for each $A \in A_i$ $(i=0,\ldots,n)$ and denote the set of pairs (α,β) with $\alpha < \beta$ and $c_i^{\alpha\beta} = c_i^{\beta\alpha} \neq 0$ by E_i , its elements by E . Then we have in $\hat{T}_\Phi(D_\Sigma^{loc}) = \prod_{\alpha=1}^{N} T_{\Phi^\alpha}(D_{\Sigma^\alpha})$ the relation

$$\sum_{i=0}^{n} \sum_{A\in A_i} x_i^A \, d\, p_i^{\alpha_A} = \sum_{i=0}^{n} \sum_{(\alpha,\beta)=E\in E_i} K_i^E (dp_i^\alpha - dp_i^\beta)$$

for some appropriate K_i^E . For $E = (\alpha,\beta)$ and $\gamma = 1,2,\ldots,N$ we define $\varepsilon(\gamma,E)$ by

$$\varepsilon(\gamma,E) = \delta_\gamma^\alpha - \delta_\gamma^\beta = \begin{cases} 0 & \text{if } \gamma \neq \alpha,\beta \\ 1 & \text{if } \gamma = \alpha \\ -1 & \text{if } \gamma = \beta \end{cases} .$$

Then we rewrite the above equation in the form

$$\sum_{i=0}^{n} \sum_{A\in A_i} x_i^A dp_i^{\alpha_A} = \sum_{i=0}^{n} \sum_{E\in E_i} K_i^E \sum_{\alpha=1}^{N} \varepsilon(\alpha,E) dp_i^{\alpha}$$

which in turn is equivalent to

$$\sum_{i=0}^{n} \sum_{\alpha=1}^{N} \Big(\sum_{A\in A_i} \delta_{\alpha}^{\alpha_A} x_i^A - \sum_{E\in E_i} K_i^E \cdot \varepsilon(\alpha,E)\Big) dp_i^{\alpha} = 0 .$$

It now follows from the discussion in § 1, that there exist λ_α $(\alpha=1,2,\ldots,N)$ with

$$\sum_{A\in A_i} \delta_{\alpha}^{\alpha_A} x_i^A - \sum_{E\in E_i} K_i^E \cdot \varepsilon(\alpha,E) = \lambda_\alpha \cdot X_i(\Phi)$$

for all α and i .

But $$\sum_{\alpha\in A} \sum_{E\in E_i} K_i^E \cdot \varepsilon(\alpha,E) = \sum_{E\in E_i} K_i^E \sum_{\alpha\in A} \varepsilon(\alpha,E) = 0$$

for all $A \in A_i$. Thus summing up the above equation over all $\alpha \in A$ for each $A \in A_i$ we get

$$x_i^A = \sum_{\alpha\in A} \lambda_\alpha X_i^{\alpha}(\Phi) = r_\Phi((\lambda_\alpha)_{\alpha=1,\ldots,N})$$

q.e.d.

The exactness at $\hat{T}_\Phi(D_\Sigma)$ and $\hat{T}_\Phi(e^{\mathbb{R}^N}\cdot\Phi)$ follows similarly from the discussion in § 1: it is equivalent to the exactness of

$$0 \to \langle dp_i^{\alpha} \mid i=0,\ldots,n;\ \alpha=1,\ldots,N\rangle \to \hat{T}_\Phi(D_\Sigma^{loc}) \to \hat{T}_\Phi(e^{\mathbb{R}^N}\cdot\Phi) \to 0$$

which follows by summing up over all $\alpha = 1,2,\ldots,N$ the exact sequences

$$0 \to \langle dp_i^{\alpha} \mid i=0,\ldots,n\rangle \to \hat{T}_{\Phi^\alpha}(D_{\Sigma^\alpha}) \to \hat{T}_\Phi(e^{\mathbb{R}}\cdot\Phi) \to 0 .$$

Proof of Theorem 4: The injectivity of $\hat{T}_{\sigma(\Phi)}(\prod_{i=0}^{n} \mathbb{R}^{A_i}) \to T_\Phi(D_\Sigma)$: It is obvious that $\hat{T}_{\sigma(\Phi)}(\prod_{i=0}^{n} \mathbb{R}^{A_i}) \to T_\Phi(D_\Sigma^{loc})$ is injective and its image is

$$\langle dx_i^A = \sum_{\alpha\in A} dx_i^{\alpha} \mid i=0,1,\ldots,n;\ A \in A_i\rangle .$$

Thus it is enough to show that

$$\langle dx_i^A \mid i=0,..,n; A \in A_i\rangle \cap \langle dp_i^{\alpha} - dp_i^{\beta} \mid i=0,1,\ldots,n; c_i^{\alpha\beta} \neq 0\rangle = 0 .$$

So - using the notations from above - let

$$d = \sum_{i=0}^{n} \sum_{E\in E_i} K_i^E \sum_{\alpha=1}^{N} \varepsilon(\alpha,E) dp_i^{\alpha} = \sum_{i=0}^{n} \sum_{A\in A_i} \mu_i^A dx_i^A = \sum_{i=0}^{n} \sum_{A\in A_i} \mu_i^A \sum_{\alpha\in A} dx_i^{\alpha}$$

represent an element in the intersection. The left hand side equals

$$\sum_{i=0}^{n} \sum_{E\in E_i} K_i^E \sum_{\alpha=1}^{N} \varepsilon(\alpha,E) \sum_{j=1}^{n} \frac{\partial^2 s}{\partial x_i \partial x_j} \, d\, x_j^\alpha \quad .$$

Thus it follows from the linear independence of the dx_j^α in $\hat{T}_\Phi(D_\Sigma^{loc})$ that

$$\mu_j^A = \sum_{i=0}^{n} \sum_{E\in E_i} K_i^E \, \varepsilon(\alpha,E) \frac{\partial^2 s}{\partial x_i \partial x_j}$$

for all $j=0,\ldots,n$ and all $\alpha \in A \in A_j$. This in turn implies

$$\sum_{i=0}^{n} \sum_{E\in E_i} \sum_{\alpha\in A} K_i^E \, \varepsilon(\alpha,E) \frac{\partial^2 s}{\partial x_i \partial x_j} \varepsilon(\alpha,E) = \sum_{\alpha\in A} \varepsilon(\alpha,E)\mu_j^A = 0$$

for all $j=0,\ldots,n$; $A \in A_j$ and $F \in E_j$. Multiplying with K_j^F and summing up over all j and $F \in E_j$ leads to

$$\sum_{\alpha=1} \sum_{i,j=0} \Big(\sum_{E\in E_i} K_i^E \, \varepsilon(\alpha,E)\Big)\Big(\sum_{F\in E_j} K_j^F \, \varepsilon(\alpha,F)\Big) \frac{\partial^2 s}{\partial x_i \partial x_j} = 0 \ .$$

It now follows from our assumption that each of the matrices $\left(\frac{\partial^2 s}{\partial x_i \partial x_j}\right)$ is negative semi-definite of maximal rank that

$$\sum_{E\in E_i} K_i^E \, \varepsilon(\alpha,E) = \lambda_\alpha X_i^\alpha(\Phi)$$

for all i and α with some appropriate $(\lambda_1,\lambda_2,\ldots,\lambda_N)$. But this implies for the left hand side of the above equation for d :

$$d = \sum_{i=0}^{n} \sum_{E\in E_i} \sum_{\alpha=1}^{N} K_i^E \, \varepsilon(\alpha,E) \, dp_i^\alpha = \sum_{i=0}^{n} \sum_{\alpha=1}^{N} \lambda_\alpha X_i^\alpha(\Phi) dp_i^\alpha = 0 \quad ,$$

q.e.d.

The surjectivity of $T_\Phi(D_\Sigma) \to T_\Phi(e^{Pl_\Phi}\cdot\Phi)$ is obvious from Theorem 3, since $e^{Pl_\Phi}\cdot\Phi$ is a submanifold of $e^{\mathbb{R}^N}\cdot\Phi$. It is also obvious that the composition

$$T_{\sigma(\Phi)}\Big(\prod_{i=0}^{n} \mathbb{R}^{A_i}\Big) \to T_\Phi(D_\Sigma) \to T_\Phi(e^{Pl_\Phi}\cdot\Phi)$$

is the zero-map: for $(\lambda_\alpha)_{\alpha=1,2,\ldots,n} \in Pl_\Phi$ we get for $dx_i^A = \sum_{\alpha\in A} dx_i^\alpha$ $(i=0,\ldots,n;$ $A \in A_i)$ evaluated at the curve

$$\Phi(\varepsilon) = (e^{\varepsilon\lambda_\alpha}\cdot\phi^\alpha)_{\alpha=1,\ldots,n} \subseteq D_\Sigma$$

$(\varepsilon \in \mathbb{R})$ equal to

$$\sum_{\alpha\in A} \frac{d}{d\varepsilon} \left(X_i^\alpha(\Phi(\varepsilon))\right)\Big|_{\varepsilon=0} = \sum_{\alpha\in A} \frac{d}{d\varepsilon} \left(e^{\varepsilon\lambda_\alpha} X_i^\alpha(\Phi)\right)\Big|_{\varepsilon=0} = \sum_{\alpha\in A} \lambda_\alpha X_i^\alpha(\Phi) = 0 \ .$$

The exactness at $\hat{T}_\Phi(D_\Sigma)$ now follows from the relation

$$\mathrm{Dim}\ T_\Phi(D_\Sigma) = \sum_{i=0}^{n} \# A_i + \mathrm{Dim}\ Pl_\Phi \quad ,$$

which is a consequence of Theorem 3.

As a consequence of Theorem 3 and Theorem 4 we get that indeed the set $D_\Sigma^o = \{\Phi \in D_\Sigma \mid Pl_\Phi = 0\}$ is a submanifold of $\mathbb{R}^{(n+1)N}$ of dimension $\sum_{i=0}^{n} \# A_i$ and that

$$D_\Sigma^o \to \prod_{i=0}^{n} \mathbb{R}^{A_i} : \Phi \longrightarrow \Big(\sum_{\alpha\in A} X_i^\alpha(\Phi)\Big)_{i=0,\dots,n}^{A\in A_i}$$

is a local diffeomorphism.

Moreover, it follows that the subspace of $\hat{T}\Phi(D\Sigma)$, generated by the differentials df of intensive functions $f : D_\Sigma \to \mathbb{R}$, which by its definition is the kernel of $T_\Phi(D_\Sigma) \to \hat{T}_\Phi(e^{\mathbb{R}^N}\cdot\Phi)$, is equal to the subspace generated by the $\langle dp_i^A \mid i=0,\dots,n;\ A \in A_i\rangle$.

We are now ready to point out the mathematical difference of simple and composite thermodynamic systems: if Σ is a "well defined" composite system, i. e. a system without free play, we can write its fundamental equation in the form

$$S(\Phi) = \sum_{\alpha=1}^{N} S^\alpha(\Phi) = \sum_{\alpha=1}^{N}\sum_{i=0}^{n} p_i^\alpha(\Phi)\cdot X_i^\alpha(\Phi) = \sum_{i=0}^{n}\sum_{A\in A_i} p_i^A(\Phi)\cdot X_i(\Phi)$$

and we have

$$dS = \sum_{\alpha=1}^{N} dS^\alpha = \sum_{\alpha=1}^{N}\sum_{i=0}^{n} p_i^\alpha(\Phi)dX_i^\alpha = \sum_{i=0}^{n}\sum_{A\in A} p_i^A(\Phi)dX_i^A \quad .$$

Thus we have - as it should be - $\frac{\partial S}{\partial x_i^A} = p_i^A(\Phi)$, so in that respect there is indeed no difference between simple and composite systems. But if we consider the relation between the dp_i^A , we see that for each $\alpha = 1,2,\dots,N$ we have the relation

$$0 = \sum_{i=0}^{n} X_i^\alpha(\Phi)dp_i = \sum_{i=0}^{n}\sum_{A\in A_i} \delta_\alpha^A\cdot X_i^\alpha(\Phi)\cdot dp_i^A$$

with $\delta_\alpha^A = 1$ for $\alpha \in A$ and $\delta_\alpha^A = 0$ for $\alpha \notin A$. According to Theorem 3 these relations are independent, if Φ has no play, so we get the following result:
if $s = s(x_o,\dots,x_m)$ is the fundamental equation of a composite system without free play, written with respect to the relevant extensive variables, then the number N

of subsystems involved can be read of from the rank of $\frac{\partial^2 s}{\partial x_i \partial x_j}$, i. e. we have

$$m = N + \text{rank}\left(\frac{\partial^2 s}{\partial x_i \partial x_j}\right) .$$

Correspondingly, the dimension of the space of parameter systems $(\mu_i)_{i=0,..,m} \in \mathbb{R}^{m+1}$ with $\sum_{i=0}^{m} \mu_i dp_i$ will be equal to N, i. e. there is more than one Gibbs-Duhem relation and altogether there are as many linearly independent ones as there are well-defined subsystems.

This corresponds well to the fact that even for simple thermodynamic systems Σ the hessian $\left(\frac{\partial^2 s_\Sigma}{\partial x_i \partial x_j}\right)$ has maximal rank only for those equilibrium states which are in a one-phase region and that at coexistence points or curves, faces or hyperfaces the rank drops according to the number of coexisting phases, which in turn is closely related to the Gibbs phase rule.

It should finally be remarked that what Callen [3] calls the basic problem of thermodynamics is closely related to the analysis of the local diffeomorphism

$$D^o_\Sigma \to \prod_{i=0}^{n} \mathbb{R}^{A_i} : \Phi \longrightarrow \Big(\sum_{\alpha \in A} X_i^\alpha(\Phi) \Big)^{A \in A_i}_{i=0,\dots,n}$$

and the construction of its (local) inverse. In fact, much of what has been worked out in this appendix can be viewed as a mathematicians comment to some of the ideas pointed out in Callen's book.

COMPLETENESS OF THREE BODY QUANTUM SCATTERING

Volker Enss

Mathematisches Institut
Ruhr-Universität

D-4630 Bochum 1
Fed. Rep. Germany

Abstract.

We outline a time-dependent proof of asymptotic completeness for scattering of three quantum mechanical particles which may be distinguishable or identical. The particles interact with pair potentials of short range, i.e. roughly with decay like $r^{-1-\varepsilon}$, $\varepsilon>0$ towards infinity. In particular the two body subsystems may have bound states or resonances at zero energy, and the three body system may have infinitely many scattering channels. The dimension is arbitrary.

Contents

I. Introduction

Consider a system of three quantum particles which interact among each other by forces depending only on the relative positions (and momenta); exterior forces are absent. Physical intuition and experience suggests that the asymptotic time evolution of such a system with forces from a suitable class should be described completely by the following alternatives:

(a) all three particles are bounded forever;

(b) a two body subsystem is in a bound state, the third particle moves freely relative to the two particle cluster in the far future;

(c) all three particles are moving freely relative to each other in the far future.

A general state should be a superposition of states with this asymptotic behaviour in the future.An analogous (in general different) decomposition should be possible for the past.

For two body systems asymptotic completeness is established by a wide variety of methods (see e.g.[20]). There the alternative simplifies to (a'): the particles are bounded forever, (b'): in the far future and remote past the two particles move freely relative to each other. The states with the asymptotic behaviour a' (resp. b') are those which lie in the space spanned by the eigenvectors (resp., in the continuous spectral subspace) of the Hamiltonian. For three body systems Ruelle's theorem [22,1] applies as well. It says that states with behaviour (a) are exactly those in the point spectral subspace of the Hamiltonian H. It remains to show that the states from the continuous spectral subspace of H (= the orthogonal complement of the eigenvectors) asymptotically decay into two or more subsystems which move freely relative to each other, i.e. that they obey (b) or (c). (As a byproduct one gets that the Hamiltonian does not have a singular continuous spectrum.)

Our strategy to prove this exploits our physical intuition. In a first step one determines subsets of the state space, the elements of which have a future time evolution well approximated by a simpler one: If two particles are in a bound state, the third is far and running away from them, then in the future the third particle will move freely relative to the pair, the forces can be neglected which couple the third particle to those in the pair. The same applies if the pair is in a low energy state, the relative velocities are small, whereas the third particle is far and running away with a higher speed. Finally, if all three particles are far separated and any two of them are running away from each other, then the future time evolution will be totally free, all forces can be neglected. To show these facts one uses that one has good control of propagation properties of states under the free time evolution. The approximations become better, if the separation of the particles at the starting time is larger.

The other part of the proof is to show that the subsets with an approximately simple future time evolution are "absorbing". This means that the component of the continuum state which does not lie in one of these subsets will be arbitrarily small if one waits long enough. This is accomplished by looking at suitable observables. Their asymptotic time evolution is simple and independent of the interaction. Therefore we can control it for the true time evolution. We will discuss the observables and their evolution in detail in Section VI.

The proof will be given here for pair potentials with a decay like $r^{-1-\varepsilon}$, $\varepsilon>0$. To simplify the presentation and to make the main ideas more transparent we will assume in this report more about the potentials than necessary. This avoids some purely technical arguments. The full proof for the more general situation will be given in [11]. If Coulomb forces are present one has to use modified free time evolutions, the treatment of these systems will be given elsewhere.

Our method of proof is geometrical and time dependent as introduced for the two body problem in [4]. Extensions, refinements, and variants of the method (short range) were given in [2,3,7,13,16,18,23,24]. The first version of the proof for two body systems as given in [7, Section IX] is closest to the present one.

Of the earlier work on completeness for three body systems we mention in particular the pioneering work of Faddeev [12] and its extensions by Ginibre and Moulin [14] and Merkuriev [15] (see e.g. [11] for further references.) The main limitation of that approach is that the short range forces have to decay roughly like $r^{-2-\varepsilon}$, $\varepsilon>0$, and that the two body subsystems should not have zero energy bound states or resonances. With an extremely complicated chain of estimates Merkuriev is able to add Coulomb forces exploiting the special features of that potential. Our approach is very different from theirs.

Very general results on absence of a singular continuous spectrum of the Hamiltonian have been given recently by Mourre [17] and Perry, Sigal, Simon [19]. That approach is related to ours in spirit but different in the details.

Before giving a precise mathematical statement of our results in

Section III we have to introduce some notation.

II. Notation and Assumptions

We consider three particles with masses m_i moving in ν-dimensional space. As usual we separate off the free center of mass motion. We use Jacobi coordinates to describe the remaining degrees of freedom. With α we label the three possibilities to pick a pair of particles $(i,j), i<j$ and the third particle k. Denote by $x_\alpha = x_i - x_j$ the relative position in the pair and by $y_\alpha = x_k - (m_i+m_j)^{-1}(m_i x_i + m_j x_j)$ the position of the third particle relative to the center of mass of the pair. The state space $\mathcal{H}$ can be represented as $L^2(\mathbb{R}^{2\nu}, d^\nu x_\alpha \, d^\nu y_\alpha)$ for any decomposition α. The canonically conjugate operators $p_\alpha = -i\,\partial/\partial x_\alpha$ and $q_\alpha = -i\,\partial/\partial y_\alpha$ are the momentum operators for the internal motion of the pair and for the motion of the third particle realtive to the pair, respectively. With the reduced masses $\mu_\alpha = m_i m_j (m_i+m_j)^{-1}$ for the pair and $\nu_\alpha = m_k(m_i+m_j)(m_i+m_j+m_k)^{-1}$ for the third particle the kinetic energy $\overline{H}_o = \Sigma(2m_k)^{-1} p_k^2$ restricted to $\Sigma\, p_k = 0$ is obtained as

$$H_o = (2\mu_\alpha)^{-1}\, p_\alpha^2 + (2\nu_\alpha)^{-1}\, q_\alpha^2 \qquad (2.1)$$

$$= \quad h_o^\alpha \quad + \quad k_o^\alpha \; .$$

Note that in the representation

$$\mathcal{H} = L^2(\mathbb{R}^{2\nu}, d^\nu x_\alpha \, d^\nu y_\alpha) = L^2(\mathbb{R}^\nu, d^\nu x_\alpha) \otimes L^2(\mathbb{R}^\nu, d^\nu y_\alpha) \qquad (2.2)$$

the kinetic energy operators have a product structure

$$h_o^\alpha = \hat{h}_o^\alpha \otimes \mathbb{1}_{y_\alpha}, \; k_o^\alpha = \mathbb{1}_{x_\alpha} \otimes \hat{k}_o^\alpha \; . \qquad (2.3)$$

The interaction is described by pair potentials which couple the particles in the pair

$$V_\alpha = V_\alpha(x_\alpha) = \hat{V}_\alpha(x_\alpha) \otimes \mathbb{1}_{y_\alpha} \; . \qquad (2.4)$$

The channel Hamiltonians are

$$H_\alpha = H_o + V_\alpha = (\hat{h}_o^\alpha + \hat{V}_\alpha) \otimes \mathbb{1}_{y_\alpha} + \mathbb{1}_{x_\alpha} \otimes \hat{k}_o^\alpha \qquad (2.5)$$

$$= \quad h^\alpha \quad + \quad k_o^\alpha \; .$$

Selfadjoint H_α generate the unitary channel-time evolution group

$$\exp(-iH_\alpha t) = \exp(-i\hat{h}^\alpha t) \otimes \exp(-i\hat{k}_o^\alpha t) . \qquad (2.6)$$

In the full Hamiltonian H all particles interact

$$H=H_o+\sum_\alpha V_\alpha \ . \tag{2.7}$$

We use freely the functional calculus to define functions of self-adjoint operators, in particular $F(\cdot)$ denotes a spectral projection of the self-adjoint operator and the indicated region. Some of them have a natural product structure. E.g.

$$F(|x_\alpha|<\rho)\ F(|y_\alpha|>R)=\hat{F}(|x_\alpha|<\rho)\ \otimes\ \hat{F}(|y_\alpha|>R)$$

is the multiplication operator with the characteristic function of the given region.

In the following we will always omit the $\hat{}$ since it will be clear from the context on which space the operators act.

To simplify the presentation we will assume here that the pair potentials V_α are multiplication operators with continuously differentiable functions $V_\alpha(x_\alpha)$ such that

$$|x_\alpha|^{1+\varepsilon} V_\alpha(x_\alpha) \text{ is bounded,} \tag{2.8}$$

$$\vec{x}_\alpha\cdot(\vec{\nabla}V_\alpha)(x_\alpha) \to 0 \text{ as } |x_\alpha| \to \infty, \text{ and it is bounded.} \tag{2.9}$$

With some care in technical questions essentially the same proof extends to a wide class of unbounded short range potentials [11].

The Hamiltonians H_o, H_α, H are all self-adjoint on a common domain which is left invariant by all time evolutions in question, and the operators are bounded relative to each other.
The same applies to h_o^α and h^α.

For any Hamiltonian we denote by $P^p(\cdot)$ and $P^{cont}(\cdot)$ the projections to the point- and continuous spectral subspaces $\mathcal{H}^p(\cdot)$ and $\mathcal{H}^{cont}(\cdot)$.

For each of the three decompositions α the corresponding two cluster wave operator is

$$\Omega_{\mp}^{\alpha} = \text{s-}\lim_{t\to\pm\infty} e^{iHt}\, e^{-iH^\alpha t}\, P^p(h^\alpha), \tag{2.10}$$

and for the totally free channel

$$\Omega_{\mp}^{o} = \text{s-}\lim_{t\to\pm\infty} e^{iHt}\, e^{-iH_o t} \ . \tag{2.11}$$

It is well known (see e.g. Theorem XI.34 in [20] that in $\nu\geq 3$ dimensions the wave operators exist under our assumptions. For $\nu=1,2$ additional assumptions about the decay rate of eigenfunctions for two body subsystems allow to prove existence as was done implicitly in [6]. The existence of the wave operators in any dimension without additional assumptions is a byproduct of the present approach. They have pairwise orthogonal ranges, which are contained in the absolutely continuous spectral subspace $\mathcal{H}^{ac}(H)$. Each of the ranges is invariant under $\exp(-iHt)$.

III. Main Results

Now we are prepared to give a precise mathematical statement of asymptotic completeness (in its strongest form). Let $\gamma \in \{0,\alpha\}$, i.e. γ labels a pairing α or the splitting into three separate particles. Asymptotic completeness is the pair of statements

$$\bigoplus_{\gamma} \operatorname{Ran} \Omega_{\pm}^{\gamma} = \mathcal{H}^{ac}(H), \tag{3.1}$$

$$\mathcal{H}^{cont}(H) = \mathcal{H}^{ac}(H). \tag{3.2}$$

The second statement says that H has no singular continuous spectrum. We will give an equivalent statement with a direct intuitive physical interpretation:

For any $\Psi \in \mathcal{H}^{cont}(H)$, $\varepsilon>0$, there is a (sufficiently late) time τ and a splitting $\Phi^{\gamma}(\tau)$ such that

$$\| e^{-iH\tau} \Psi - \sum_{\gamma} \Phi^{\gamma}(\tau) \| < \varepsilon \tag{3.3}$$

and that for all γ

$$\sup_{t\geq 0} \| (e^{-iHt} - e^{-iH^{\gamma}t}) \Phi^{\gamma}(\tau) \| < \varepsilon \ . \tag{3.4}$$

Similarly for the past.

This means that any state orthogonal to the bound states after a long time splits into (at least) two subsystems which approximately do not interact in the further future. One can omit the term with $\gamma=0$ in the finite sum, since there are no point spectral projections in (3.4), but it is convenient to list it as well. We will show the equivalence of the two statements below.

Our main result is

Theorem 3.1. Let the Hamiltonian H of a quantum mechanical three body system satisfy (2.7)-(2.9), then asymptotic completeness holds, and H has no singular continuous spectrum.

Remark: The theorem is proved under much weaker assumptions in [11].

Corollary 3.2. Theorem 3.1 remains true if two or all three particles are indistinguishable.

Proof. For identical particles the physical state space consists of certain (anti-)symmetric wave functions. The decompositions (3.3) may lead to unphysical components $\Phi^\gamma(\tau)$. Let the interchange of two identical particles transform a decomposition α into α'. Our procedure to pick from $\exp(-iH\tau)\Psi$ the component $\Phi^\alpha(\tau)$ (as described in Section VII) differs from that to pick $\Phi^{\alpha'}(\tau)$ just by permutation of the identical particles. Similarly H^α is transformed into $H^{\alpha'}$. Therefore

$$\sum_\gamma e^{-iH^\gamma t}\,\Phi^\gamma(\tau) \tag{3.5}$$

is in the physical state space again. It obeys for any $\varepsilon>0$, τ large enough

$$\sup_{t\geq 0}\|e^{-iH(t+\tau)}\,\Psi - \sum_\gamma e^{-iH^\gamma t}\,\Phi^\gamma(\tau)\|<4\varepsilon. \tag{3.6}$$

Thus asymptotic completeness holds by Lemma 3.3.

□

We will show the equivalence of our different descriptions of asymptotic completeness for a more general class of Hamiltonians.

Lemma 3.3. Assume that for all two body subsystems the Hamiltonians h^α are asymptotically complete and that the three body wave operators exist. Then for the three body system the notions of asymptotic completeness (3.1 & 2), (3.3 & 4), and (3.3 & 6) are equivalent.

Proof. Assume (3.1 & 2), $\Psi \in \mathcal{H}^{cont}(H)$ can be decomposed $\Psi = \sum \Psi^\gamma$, $\Psi^\gamma \in \mathrm{Ran}(\Omega_-^\gamma)$. Then there exist $\Phi^\alpha \in \mathcal{H}^p(h^\alpha)$, $\Phi^0 \in \mathcal{H}$ such that $\lim_{t\to\infty}\|\exp(-iHt)\,\Psi^\gamma - \exp(-iH^\gamma t)\,\Phi^\gamma\|=0$. In particular there is a τ such that for all γ

$$\sup_{t\geq 0}\|e^{-iH(t+\tau)}\,\Psi^\gamma - e^{-iH^\gamma(t+\tau)}\,\Phi^\gamma\| < \frac{\varepsilon}{2}\,. \tag{3.7}$$

Set $\Phi^\gamma(\tau)=\exp(-iH\tau)\,\Psi^\gamma$, it satisfies (3.3) and (3.4) by

$$\begin{aligned}
&\|(e^{-iHt} - e^{-iH^\gamma t})\,\Phi^\gamma(\tau)\|\\
&\leq\|e^{-iH(t+\tau)}\Psi^\gamma - e^{-iH^\gamma(t+\tau)}\,\Phi^\gamma\|\\
&+\|e^{-iH^\gamma\tau}\,\Phi^\gamma - e^{-iH\tau}\,\Psi^\gamma\|\\
&<\varepsilon \text{ for all } t\geq 0.
\end{aligned}$$

Obviously (3.6) follows from (3.4).

Finally for a given $\varepsilon>0$ fix τ and the decomposition (3.3) such that (3.6) holds. Decompose each $\Phi^{\alpha}(\tau) = P^{p}(h^{\alpha})\ \Phi^{\alpha}(\tau) + P^{cont}(h^{\alpha})\ \Phi^{\alpha}(\tau)$.

Completeness of the two body subsystems implies that there is an $s<\infty$ such that for all α

$$\sup_{t\geq 0} \| (e^{-iH_0 t} - e^{-iH^{\alpha}t})\ e^{-iH^{\alpha}s}\ P^{cont}(h^{\alpha})\ \Phi^{\alpha}(\tau)\| < \varepsilon\ . \tag{3.8}$$

At time $\tau'=\tau+s$ we regroup the terms to

$$\begin{aligned} \Psi^{\alpha}(\tau') &= P^{p}(h^{\alpha})\ e^{-iH^{\alpha}s}\ \Phi^{\alpha}(\tau) \\ \Psi^{0}(\tau') &= \sum_{\alpha} e^{-iH^{\alpha}s}\ P^{cont}(h^{\alpha})\ \Phi^{\alpha}(\tau) + \\ &\quad + e^{-iH_0 s}\ \Phi^{0}(\tau)\ . \end{aligned} \tag{3.9}$$

Then

$$\sup_{t\geq 0} \| e^{-iH(t+\tau')}\ \Psi - \sum_{\gamma} e^{-iH^{\gamma}t}\ \Psi^{\gamma}(\tau')\| < 7\varepsilon\ .$$

Since $\lim_{t\to\infty} e^{iHt}\ e^{-iH^{\gamma}t}\ \Psi^{\gamma}(\tau') = \Omega_{-}^{\gamma}\ \Psi^{\gamma}(\tau')$ exists for any γ and τ' we get that for any $\varepsilon>0$

$$\| e^{-iH\tau'}\ \Psi - \sum_{\gamma} \Omega_{-}^{\gamma}\ \Psi^{\gamma}(\tau')\| \leq 7\varepsilon\ .$$

This implies $\Psi \in \bigoplus_{\gamma} \mathrm{Ran}\ \Omega_{-}^{\gamma}$, similarly for the other time direction. Thus (3.1 & 2) hold.

□

IV. Information about Two Body Subsystems

For our potentials one knows [21] that the strictly negative eigenvalues of the subsystem Hamiltonians h^{α} are at most finitely degenerate and that they can accumulate only at zero. Moreover it is convenient to use that there are no bound states with strictly positive energy (e.g. by Theorem XIII.58 in [21]). On the other hand it is irrelevant for our approach whether there are finitely or infinitely many bound states (corresponding to the number of scattering channels). At zero energy there may be resonances or bound states with arbitrary degeneracy.

These facts can be used to construct a convenient dense set of states in $\mathcal{H}^{cont}(H)$ for which we will verify (3.3) and (3.4). Certainly this implies that (3.3 & 4) hold on all of $\mathcal{H}^{cont}(H)$ because only bounded operators are involved. For a given pair $0<a<b<\infty$ we define the set

$$A=\{\omega \in \mathbb{R} \,|\, |\omega| > 2a,\ |\omega-E_i^\alpha| > 2a,\ \omega<b/2\} \tag{4.1}$$

where E_i^α are the finitely many eigenvalues of h^α below $-2a$ (thresholds). The set of spectral projections

$$F(H \in A) = \int_A dE^H(\lambda) \tag{4.2}$$

has a dense range in $\mathcal{H}^{cont}(H)$ if a and b vary. The states in $F(H \in A)\mathcal{H}^{cont}(H)$ have the following physical properties: If two particles are in a bound state and the third is far away, then its kinetic energy relative to the pair is at least 2a. (H and H^α approximately coincide for large separation.) Similarly if a pair has very small relative energy below a', then the third particle has minimal kinetic energy 2a-a' if it is far separated from both. We say that such states have bounded energy strictly away from thresholds.

Suppose that for a given decomposition α h^α has infinitely many bound states in $L^2(\mathbb{R}^\nu, d^\nu x_\alpha)$. Let $P^p(h^\alpha) = \sum_{i=1}^{\infty} P_i^p(h^\alpha)$ where $P_i^p(h^\alpha)$ are the eigenprojections of an orthonormal basis of eigenvectors of h^α in $\mathcal{H}^p(h^\alpha)$. For all i one has $F(h^\alpha \leq 0)\, P_i^p(h^\alpha) = P_i^p(h^\alpha)$. If one is interested in a finite region of space $|x_\alpha| < \rho < \infty$ only, one may restrict the summation over i to a finite one:

$$\| F(|x_\alpha|<\rho) \sum_{i=1}^{\infty} P_i^p(h^\alpha) - F(|x_\alpha|<\rho) \sum_{i=1}^{N} P_i^p(h^\alpha) \| \to 0 \tag{4.3}$$

as $N \to \infty$. The difference is

$$F(|x_\alpha|<\rho)\ F(h^\alpha \leq 0) \sum_{i>N} P_i^p(h^\alpha) \ .$$

It converges in norm to zero since the sum converges strongly to zero as $N \to \infty$. The first product is well known to be a compact operator for all $\rho < \infty$ ("local compactness"). Certainly (4.3) is also valid if we consider the operators as acting on $L^2(\mathbb{R}^{2\nu}, d^\nu x_\alpha\, d^\nu y_\alpha)$ with the obvious product structure.

The other important pieces of information about two body subsystems are the propagation properties in space and time. We quote from [7, Section 4] a uniform bound which extends a Theorem of Ruelle. For any $\rho, E < \infty$

$$\lim_{T\to\infty} \| \frac{1}{T} \int_0^T dt\ e^{ih^\alpha t}\, F(|x_\alpha| < \rho)\ F(h^\alpha < E)\ P^{cont}(h^\alpha)\ e^{-ih^\alpha t} \| = 0 \ . \tag{4.4}$$

This again follows from local compactness. On the three particle state space one has

$$F(h^\alpha < E)\ F(H^\alpha < E) = F(H^\alpha < E)\ , \tag{4.5}$$

therefore it follows immediately that for $\rho, E < \infty$

$$\lim_{T\to\infty} \left\| \frac{1}{T} \int_0^T dt\ e^{iH^\alpha t}\ F(|x_\alpha| < \rho)\ F(H^\alpha < E)\ P^{cont}(h^\alpha)\ e^{-iH^\alpha t} \right\| = 0\ . \tag{4.6}$$

For a given error ε we know from this that there is a time $T(\varepsilon)$ such that states from $P^{cont}(h^\alpha)$ will have $|x_\alpha| > \rho$ in the time mean. This time depends on the separation ρ and the energy cutoff E but is otherwise independent of the states.

The other propagation property does not depend on the spectral character but it reflects the fact that a particle with low energy cannot travel fast. If it is well localized initially it should stay inside a ball of growing radius $R + v_{max} \cdot t$ where v_{max} is the maximal velocity corresponding to $E_{max} = (\mu/2) v_{max}^2$. The following proposition says that for interacting particles the tails of the state in the "forbidden region" decay integrably. (For free particles this decay is faster than any inverse power of t.)

<u>Proposition 4.1.</u> On $L^2(\mathbb{R}^\nu, d^\nu x)$ let $h = h_o + V(x)$ satisfy (2.8), (2.9)

$$\lim_{R\to\infty} \int_0^\infty dt \| F(|x| > 2R+2vt)\ e^{-iht}\ \phi(h)\ F(|x| < R) \| = 0 \tag{4.7}$$

for $\phi \in C_o^\infty(\mathbb{R})$ with $\phi(e) = 0$ for $e \geq \frac{\mu}{2} v^2$.

<u>Idea of Proof.</u> The full proof is lengthy, we will restrict ourselves here to an outline of the main ideas. Full details for a wide class of potentials, including very singular ones and long range potentials (Coulomb), can be found in [10]. By Lemma 2 in [5] we know that

$$\| [\phi(h) - \phi(h_o)]\ F(|x| > r) \| \leq \frac{\text{const.}}{(1+r)^{1+\varepsilon}}\ . \tag{4.8}$$

Thus it is sufficient to estimate

$$\| F(|x| > 2R+2vt)\ \phi(h_o)\ e^{-iht}\ \phi(h)\ F(|x| < R) \| . \tag{4.8}$$

With the free time evolution instead of the interacting one we can use the propagation properties of the free time evolution. We neglect the tails which have to be estimated separately.

$$F(|x| > 2R+2vt)\ \phi(h_o)\ e^{-ih_o t}$$

$$\approx F(|x|>2R+2vt)\ \phi(h_o)\ e^{-ih_o t}\ F(|x|>2R+vt) \tag{4.9}$$

because the energy support of $\phi(h_o)$ corresponds to the maximal velocity v. This gives an integrable contribution since

$$\begin{aligned}&\| F(|x|>2R+vt)\ \phi(h)\ F(|x|<R)\| \\ &\leq \| F(|x|>2R+vt)\ [\phi(h)-\phi(h_o)]\| \\ &+\| F(|x|>2R+vt)\ \phi(h_o)\ F(|x|<R)\| .\end{aligned} \tag{4.10}$$

The first term is integrable by (4.8), the second decays rapidly since $\phi(h_o)$ acts in x-space as convolution with a rapidly decaying bounded function. The difference of the time evolutions is expanded by the Duhamel formula

$$\begin{aligned}&\int_o^t ds \| F(|x|>2R+2vt)\ \phi(h_o)\ e^{-ih_o(t-s)}\ V\ F(|x|>2R+vt+vs)\| \times \\ &\times \| F(|x|>2R+2vs)\ e^{-ihs}\ \phi(h)\ F(|x|<R)\|\end{aligned} \tag{4.11}$$

and we have used the propagation properties of the free evolution to insert the factor $F(|x|>2R+vt+vs)$ (again neglecting tails). With the shorthands

$$b(R,t)=\| F(|x|>2R+2vt)\ e^{-iht}\ \phi(h)\ F(|x|<R)\| , \tag{4.12}$$

$$K(R,t,s)=\| V\ F(|x|>2R+vt+vs)\| \leq \frac{\text{const.}}{(R+vt+vs)^{1+\varepsilon}} , \tag{4.13}$$

we have "shown"

$$b(R,t)\leq \frac{\text{const.}}{(R+vt)^{1+\varepsilon}} + \int_o^t ds\ K(R,t,s)\ b(R,s) . \tag{4.14}$$

This implies

$$\begin{aligned}(R+vt)^{1+\varepsilon}\ b(R,t)\leq \text{const} + (R+vt)^{1+\varepsilon} \int_o^t ds\ K(R,t,s)\ (R+vs)^{-1-\varepsilon} \times \\ \times (R+vs)^{1+\varepsilon}\ b(R,s).\end{aligned} \tag{4.15}$$

By Gronwall's inequality

$$(R+vt)^{1+\varepsilon}\ b(R,t)\leq \text{const.}\exp\{ \int_o^t ds\ (R+vt)^{1+\varepsilon}\ K(R,t,s)\ (R+vs)^{-1-\varepsilon}\}. \tag{4.16}$$

By (4.13) the ecponent is integrable which implies that also $b(R,t)$ is integrable and the integral is arbitrarily small for large R. The tails which we have neglected here can be bounded by const. $(R+vt)^{-1-\varepsilon'}$, $0<\varepsilon'<\varepsilon$, which has to be added to the RHS of (4.14).

□

V. Separation of the Particles

For the states in the ranges of the two cluster wave operators condition (3.4) certainly holds, we do not have to prove anything for them. Their complement in the continuous spectral subspace of H should consist only of states in which all particles move independently in the far future. In particular the particles should be far *separated* from each other. We introduce the shorthand

$$\mathcal{H}_s = \mathcal{H}^{cont}(H) \cap \Big(\bigoplus_{\alpha=1}^{3} \text{Ran}\, \Omega_-^{\alpha} \Big)^{\perp} . \tag{5.1}$$

In this section we will show separation in the time mean.

Proposition 5.1. Let $\Psi \in \mathcal{H}_s$, $F(H \in A)\,\Psi = \Psi$ as given in (4.1 & 2). Then for any $\rho < \infty$

$$\lim_{T\to\infty} \frac{1}{T} \int_0^T dt \sum_{\alpha=1}^{3} \| F(|x_\alpha| < \rho)\ e^{-iHt}\, \Psi \| = 0 . \tag{5.2}$$

Proof. We estimate each summand in (5.2) separately. It is sufficient to show that the integrand is bounded by a sum of terms which are either small in the limit $t\to\infty$ or in the time mean.

$$\begin{aligned} &\| F(|x_\alpha| < \rho)\ e^{-iHt}\, \Psi \| \\ &\leq \| F(|x_\alpha| < \rho)\ P^p(h^\alpha)\ F(|y_\alpha| > R)\ e^{-iHt}\, \Psi \| \\ &+ \| F(|x_\alpha| < \rho)\ P^{cont}(h^\alpha)\ F(|y_\alpha| > R)\ e^{-iHt}\, \Psi \| \\ &+ \| F(|x_\alpha| < \rho)\ F(|y_\alpha| < R)\ e^{-iHt}\, \Psi \| . \end{aligned} \tag{5.3}$$

The first summand is bounded uniformly in time t, it is bounded by ε for $R \geq R_1(\varepsilon)$ as shown in Lemma 5.2. For $T \geq T_2(\varepsilon)$ and $R \geq R_2(\varepsilon)$ the time average of the second term is bounded by ε following Lemma 5.4. Finally set $R = \max(R_1, R_2)$. Then $F(|x_\alpha| < \rho)\ F(|y_\alpha| < R)\ F(H \in A)$ is a compact operator in $L^2(\mathbb{R}^{2\nu})$. Since $\Psi \in \mathcal{H}^{cont}(H)$ we know by Ruelle's theorem [22,1] that there is a $T_3(\varepsilon) \geq T_2(\varepsilon)$ such that the time average of the third summand is bounded by ε for all $T \geq T_3(\varepsilon)$. Repeat the same procedure in all three decompositions α and we get a $T(\varepsilon)$ such that (5.2) is bounded by 9ε for all $T > T(\varepsilon)$.

□

It remains to show the two Lemmas used above. The first of them is essentially a two cluster estimate very similar to the two body situation. The same result has been proved in [6] for short and long range interactions. However, there we used a decay assumption (26) for the eigenfunctions of the bounded subsystem. This is no longer necessary if we use Proposition 4.1.

Lemma 5.2. Let $\Psi \in \mathcal{H}_s$, $F(H \in A)\ \Psi = \Psi$ as given in (4.1 & 2). Then for any $\rho < \infty, \alpha = 1,2,3,$

$$\lim_{R\to\infty} \sup_{t\geq 0} \| F(|x_\alpha| < \rho)\ P^p(h^\alpha)\ F(|y_\alpha| > R)\ e^{-iHt}\ \Psi \| = 0 . \tag{5.4}$$

Remark. The idea of the proof is the following: Only finitely many of the bound states contribute, the third particle has strictly positive kinetic energy relative to the bounded pair. Either it is outgoing, then for R large enough it is moving approximately freely in the future and the pair remains in the bound state in contradiction to the assumption $\Psi \in \mathcal{H}_s$. Or the third particle is incoming, then it was localized farther away from the pair initially. Since Ψ is localized somewhere this cannot be the case for very large R.

Proof. By (4.3) for a given error $\varepsilon>0$ we can estimate the norm in (5.4) uniformly in R and t by (for notational convenience we denote the cutoff by 2R instead of R in (5.4).)

$$\| \sum_{i=1}^{N} P_i^p(h^\alpha)\ F(|y_\alpha| > 2R)\ e^{-iHt}\ \Psi \| + \varepsilon , \tag{5.5}$$

the finite sum is up to $N=N(\varepsilon)$. Denote by A' the energy region strictly bigger than A which is defined as in (4.1) but with a and b replacing 2a and b/2. Then we have for all $R > R_1 = R_1(\varepsilon, N(\varepsilon))$

$$\| \sum_{i=1}^{N} P_i^p(h^\alpha)\ F(|y_\alpha| > 2R)\ \{F(H^\alpha \in A')\ F(H \in A) - F(H \in A)\} \| < \varepsilon . \tag{5.6}$$

Intuitively this is evident. If the third particle is far from both particles in the pair then the interactions coupling the third particle do not matter and the energy support is the same w.r.t. H and H_α. The bigger region A' takes care of the tails. We won't give the proof here since a more general statement has been shown under weaker assumptions in [6, Lemma 2b].

Denote by $E_{min}=\min\{E_i^\alpha\}\leq 0$ the minimal bound state energy for the pairs. On the range of $P^p(h^\alpha)\ F(H^\alpha \in A')$ the kinetic energy k_o^α of the third particle relative to the pair obeys

$$a \leq k_o^\alpha \leq b-E_{min} . \tag{5.7}$$

Therefore we can pick a function $g \in C_0^\infty(\mathbb{R})$ with $0 \leq g(e) \leq 1$, $g(e) = 1$ on $a-2a' \leq e \leq b-E_{min}$ and $g(e) = 0$ for $e \leq a-3a'$. The cutoff $0 < a' < a/3$ will be chosen later.

$$P^p(h^\alpha)\ F(H^\alpha \in A')$$

$$= P^p(h^\alpha)\ g(k_o^\alpha)\ F(H^\alpha \in A'). \tag{5.8}$$

From (5.6) we know about the norm in (5.5)

$$\| \sum_{i=1}^{N} P_i^p(h^\alpha)\ F(|y_\alpha| > 2R)\ e^{-iHt}\ \psi \|$$

$$\leq \| \sum_{i=1}^{N} P_i^p(h^\alpha)\ F(|y_\alpha| > 2R)\ g(k_o^\alpha)\ e^{-iHt}\ \psi \| + \varepsilon \tag{5.9}$$

for given N, and for all $\alpha, t, a' > 0$, and $R > R_1(\varepsilon)$. In this expression we introduce the phase space decomposition into incoming and outgoing parts for the third particle relative to the pair. With slight changes we follow [8,Section V]. Decompose $g(k_o^\alpha)$ into a finite sum

$$g(k_o^\alpha) = \sum_j \chi_j^2(\vec{q}_\alpha) \tag{5.10}$$

with each $\chi_j \in C_0^\infty(\mathbb{R}^\nu)$, $0 \leq \chi_j(\cdot) \leq 1$, and for $m = \min_\alpha \nu_\alpha$

$$\text{supp}\ \chi_j(\vec{q}) \subset \{\vec{q} \mid |\vec{q} - m\vec{v}_j| \leq m |\vec{v}_j| \sin 10^\circ\} \tag{5.11}$$

for suitably chosen $\vec{v}_j$. This gives a decomposition of a state into pieces with velocities near $\vec{v}_j$. In x-space we use a decomposition into unit cells with centers $\vec{a} \in Z^\nu$ which sums up to the identity:

$$F_{\vec{a}}\ (\vec{y}) = \prod_{i=1}^{\nu} F(|y_i - a_i| < \tfrac{1}{2}). \tag{5.11'}$$

Then we have

$$g(k_o^\alpha) = \sum_j \sum_{\vec{a} \in Z^\nu} \chi_j(\vec{q}_\alpha)\ F_{\vec{a}}(\vec{y}_\alpha)\ \chi_j(\vec{q}_\alpha). \tag{5.12}$$

In the bound (5.9) the product $F(|y_\alpha| > 2R)\ g(k_o^\alpha)$ occurs. Since each of the finitely many $\chi_j(\vec{q}_\alpha)$ act in y_α-space as convolution operators with bounded and rapidly decaying kernels the summation over $\vec{a}$ in (5.12) can be restricted to $|\vec{a}| > R$, the error is bounded in norm by ε for $R > R_2(\varepsilon) \geq R_1$. As usual we define the incoming and outgoing parts by restricting the relative directions of $\vec{a}$ and $\vec{v}_j$:

$$P_R^{out} = \sum_j \sum_{\substack{|\vec{a}|>R \\ (\vec{a}\cdot\vec{v}_j)\geq 0}} \chi_j(\vec{q}\ ) \ F_{\vec{a}}(\vec{y}_\alpha) \ \chi_j(\vec{q}_\alpha) \ , \tag{5.13}$$

$$P_R^{in} = \sum_j \sum_{\substack{|\vec{a}|>R \\ (\vec{a}\cdot\vec{v}_j)< 0}} \chi_j(\vec{q}_\alpha) \ F_{\vec{a}}(\vec{y}_\alpha) \ \chi_j(\vec{q}_\alpha) \ . \tag{5.14}$$

Both operators are positive. Using them we can replace the bound (5.9) by

$$\begin{aligned} &\| \sum_{i=1}^{N} P_i^p(h^\alpha) \ F(|y_\alpha|>2R) \ e^{-iHt} \ \Psi\| \\ &\leq \| \sum_{i=1}^{N} P_i^p(h^\alpha) \ P_R^{out} \ e^{-iHt} \ \Psi\| \\ &+\| \sum_{i=1}^{N} P_i^p(h^\alpha) \ P_R^{in} \ e^{-iHt} \ \Psi\| + 2\varepsilon \end{aligned} \tag{5.15}$$

It holds for given N, and for all $\alpha, t, a'>0$, and $R>R_2$. Due to the positivity of $\Sigma \ P_i^p(h^\alpha) \ P_R^{out}$ it is sufficient for the first summand in (5.15) to show that

$$\lim_{R\to\infty} \sup_t \ (e^{-iHt} \ \Psi, \ \Sigma \ P_i^p(h^\alpha) \ P_R^{out} \ e^{-iHt} \ \Psi) = 0 \ . \tag{5.16}$$

With $\Psi \in \mathcal{H}_s$ also $e^{-iHt} \ \Psi \in (\text{Ran} \ \Omega_-^\alpha)^\perp$ for all t. We can insert in the scalar product a factor $(1 - \Omega_-^\alpha)$ left of the summation. In the next Lemma we will show for suitably chosen $a'>0$ that

$$\lim_{R\to\infty} \| (1 - \Omega_-^\alpha) \sum_{i=1}^{N} P_i^p(h^\alpha) \ P_R^{out}\| = 0 \ , \tag{5.17}$$

thus (5.16) holds. Concerning the second summand in (5.15) the next Lemma shows that

$$\lim_{R\to\infty} \sup_{t\geq 0} \| \sum_{i=1}^{N} P_i^p(h^\alpha) \ P_R^{in} \ e^{-iHt} \ F(|y_\alpha| < \tfrac{R}{2})\| = 0 \ . \tag{5.18}$$

Using finally that

$$\lim_{R\to\infty} \|F(|y_\alpha|> R/2) \ \Psi\| = 0 \tag{5.19}$$

we have estimated all terms and Lemma 5.2 is proved. To make the norm in (5.4) smaller than ε we must choose R big enough. How big is determined by ρ, by the energy support A of Ψ and by the initial localization of Ψ as seen in (5.19). It is otherwise independent of Ψ.

□

The choice of a' is a matter of kinematics. We may assume that in the decomposition (5.10 & 11) all velocities $\vec{v}_j$ are chosen such that the corresponding kinetic energy $(\nu_\alpha/2)|v_j|^2 > a-3a'$, i.e. is in the support of g for at least one α. Thus the minimal speed of the third particle relative to the center of mass of the pair is

$$\min|\vec{v}_j| > \{2(a-3a')/\max \nu_\alpha\}^{1/2} . \tag{5.20}$$

We will allow soon that the particles in the pair have a relative kinetic energy bounded by 2a' which corresponds to a maximal speed

$$v_o = \{4a'/\min \mu_\alpha\}^{1/2} . \tag{5.21}$$

If $v_o < \min|v_j|$ we can be shure that the third particle moves away from each particle in the pair. To simplify the notation we require

$$6v_o \leq \min|v_j| \Leftrightarrow \frac{a'}{a-3a'} \leq \frac{\min \mu_\alpha}{72 \max \nu_\alpha} . \tag{5.22}$$

Since a>0 such an a' can be chosen, it depends only on the energy support A of the state Ψ. We keep it fixed in the sequel.

Next we choose a function $\phi \in C_o^\infty(\mathbb{R})$, $0 \leq \phi(e) \leq 1$ with $\phi(e)=1$ (resp. 0) for $E_{min} \leq e \leq a'$ (resp. $\geq 2a'$). With such a function we have in particular $P^p(h^\alpha) = P^p(h^\alpha)\ \phi(h^\alpha)$. States in the range of $\phi(h^\alpha)$ have energy bounded by 2a' and those in the range of $[1 - \phi(h_o^\alpha)]$ have strictly positive kinetic energy for the relative motion of the particles in the pair.

<u>Lemma 5.3.</u> For ϕ, P_R^{in}, and P_R^{out} as given above

a) $$\lim_{R\to\infty} \sup_{t\geq 0} \|[e^{-iHt} - e^{-iH^\alpha t}]\ \phi(h^\alpha)\ F(|x_\alpha|<R/6)\ P_R^{out}\| = 0, \tag{5.23}$$

b) $$\lim_{R\to\infty} \sup_{t\geq 0} \|F(|x_\alpha|<R/6)\ \phi(h^\alpha)\ P_R^{in}\ [e^{-iHt} - e^{-iH^\alpha t}]\| = 0 , \tag{5.24}$$

c) $$\lim_{R\to\infty} \sup_{t\geq 0} \|[e^{-iHt} - e^{-iH^\alpha t}] \sum_{i=1}^{N} P_i^p(h^\alpha)\ P_R^{out}\| = 0, \tag{5.25}$$

d) $\lim_{R\to\infty} \|[1-\Omega_-^\alpha] \sum_{i=1}^{N} P_i^p(h^\alpha)\ P_R^{out}\| = 0$, (5.26)

e) $\lim_{R\to\infty} \sup_{t\geq 0} \| \sum_{i=1}^{N} P_i^p(h^\alpha)\ P_R^{in}\ e^{-iHt}\ F(|y_\alpha|<R/2)\| = 0$, (5.27)

c) – e) hold for any $N < \infty$.

Proof. a) We use Cook's method to estimate (5.23):

$$\int_0^\infty dt\| (H-H^\alpha)\ e^{-iH^\alpha t}\ \phi(h^\alpha)\ F(|x_\alpha|<R/6)P_R^{out}\|$$

$$\leq \int_0^\infty dt\|H-H^\alpha\|\cdot\|F(|x_\alpha|>\tfrac{R}{3}+2v_o t)\ e^{-ih^\alpha t}\ \phi(h^\alpha)\ F(|x_\alpha|<R/6)\|$$

$$+ \int_0^\infty dt\|H-H^\alpha\|\cdot\|F(|y_\alpha|<\tfrac{R}{2}+3v_o t)\ e^{-ik_o^\alpha t}\ P_R^{out}\| \quad (5.28)$$

$$+ \int_0^\infty dt\|(H-H^\alpha)\ F(|x_\alpha|<\tfrac{R}{3}+2v_o t)\ F(|y_\alpha|>\tfrac{R}{2}+3v_o t)\| .$$

The first term is integrable and the integral vanishes as $R\to\infty$ by Proposition 4.1. The same follows for the second term from well known propagation properties for the free time evolution, see e.g. [8, Lemma 5.3] (Our slightly different definition of P_R^{out} does not affect the proof, the factor 1/4 in the cutoff can be easily improved to 1/2 in our situation.) Finally for the third summand we use the integrable decay of the potentials. Le x_k be the coordinate of the third particle and x_j for one particle in the pair, then $|x_\alpha|<R/3+2v_o t$ and $|y_\alpha|>R/2+3v_o t$ together imply $|x_j-x_k|>R/6+v_o t$. We get as a bound

$$\int_0^\infty dt \sum_{\beta\neq\alpha} \|V_\beta(x_\beta)\ F(|x_\beta|>R/6+v_o t)\| \quad (5.29)$$

which is finite by (2.8) and arbitrary small for large R. This finishes the proof of part a).

b) The adjoint operator of the one in the norm in a) is

$$F(|x_\alpha|<R/6)\ \phi(h^\alpha)\ P_R^{out}[e^{-iH(-t)} - e^{-iH^\alpha(-t)}] . \quad (5.30)$$

The transformation $t\to -t$, $p_\alpha\to -p$, $q_\alpha\to -q$, leaving the coordinates unchanges, does not affect the kinematics or dynamics. It transforms (5.30) into (5.24).

c) follows from a) and the observations that

$$\sum_{i=1}^{N} P_i^p(h^\alpha) = \phi(h\ ) \sum_{i=1}^{N} P_i^p(h^\alpha) \quad \text{and} \tag{5.31}$$

$$\lim_{R\to\infty} \| F(|x_\alpha|>R/6) \sum_{i=1}^{N} P_i^p(h^\alpha) \| = 0 \quad \text{for any } N. \tag{5.32}$$

d) is an obvious consequence of c) by taking $t\to\infty$.

e) By the adjoint and transformed estimate of c) replace $\exp(-iHt)$ by $\exp(-iH^\alpha t)$ and use that

$$\lim_{R\to\infty} \sup_{t\geq 0} \| P_R^{in}\, e^{-ik_o^\alpha t}\, F(|y_\alpha|<R/2+3v_o t) \| = 0 . \tag{5.33}$$

□

Finally we turn to

<u>Lemma 5.4.</u> For any $\rho<\infty$, $F(H\in A)\ \Psi = \Psi$, $\varepsilon>0$ there are $T(\varepsilon)$ and $R(\varepsilon)$ such that for all $T>T(\varepsilon)$, $R>R(\varepsilon)$

$$\frac{1}{T}\int_0^T dt \| F(|x_\alpha|<\rho)\ P^{cont}(h^\alpha)\ F(|y_\alpha|>R)\ e^{-iHt}\ \Psi \| < \varepsilon . \tag{5.34}$$

<u>Proof.</u> If for some $T(\varepsilon)$, R we can show for all $\tau\geq 0$

$$\frac{1}{T(\varepsilon)}\int_0^{T(\varepsilon)} dt \| F(|x_\alpha|<\rho)\ P^{cont}(h^\alpha)\ F(|y_\alpha|>R)\ e^{-iH(t+\tau)}\ \Psi \| < \frac{\varepsilon}{2} \tag{5.35}$$

then (5.34) holds for all $T\geq T(\varepsilon)$. Again from [6, Lemma 2b] we deduce analogously to (5.6) that there is an $R_1=R_1(\varepsilon,\rho)$ such that for all $R>R_1$

$$\| F(|x_\alpha|<\rho)\ P^{cont}(h^\alpha)\ F(|y_\alpha|>R)\ [F(H^\alpha<b)-\mathbb{1}]\ F(H\in A) \| < \frac{\varepsilon}{8} . \tag{5.36}$$

Determine $T(\varepsilon)$ such that the norm in (4.6) is bounded by $(\varepsilon/8)^2$ then by the Cauchy Schwarz inequality uniformly in R,τ

$$\frac{1}{T(\varepsilon)}\int_0^{T(\varepsilon)} dt \| F(|x_\alpha|<\rho)\ P^{cont}(h^\alpha)\ F(|y_\alpha|>R)\ F(H^\alpha<b)\ e^{-iH^\alpha t}\ e^{-iH\tau}\ \Psi \| < \frac{\varepsilon}{8} . \tag{5.37}$$

Keep $T(\varepsilon)$ fixed in the sequel. It remains to show that

$$\sup_{0\leq t\leq T(\varepsilon)} \| F(|x_\alpha|<\rho)\ P^{cont}(h^\alpha)\ F(|y_\alpha|>R)\ F(H^\alpha<b)\ [e^{-iH^\alpha t} - e^{-iHt}] \| < \frac{\varepsilon}{4} \tag{5.38}$$

for large enough R. On the range of $F(|x_\alpha|<\rho)\ F(|y_\alpha|>R')$ we have

$|x_\beta|>R'-\rho$ for $\beta\neq\alpha$. Therefore we can find an R' such that

$$\sum_{\beta\neq\alpha} \|F(|x_\alpha|<\rho)\; P^{cont}(h^\alpha)\; F(|y_\alpha|>R')\; V_\beta\| < \varepsilon/8\; T(\varepsilon). \tag{5.39}$$

We expand the difference in (5.38) by the Duhamel formula and obtain the bound

$$\int_0^{T(\varepsilon)} dt \sum_{\beta\neq\alpha} \{\|F(|x_\alpha|<\rho)\; P^{cont}(h^\alpha)\; F(|y_\alpha|>R)\; F(H^\alpha<b)\; e^{-iH^\alpha t}\; F(|y_\alpha|<R')\|\cdot\|V_\beta\|$$
$$+\|F(|x_\alpha|<\rho)\; P^{cont}(h^\alpha)\; F(|y_\alpha|>R')\; V_\beta\|\}. \tag{5.40}$$

The contribution from the second summand is bounded by $(\varepsilon/8)$ due to (5.39).

For the first term observe that

$$K(t) := F(|x_\alpha|<\rho)\; P^{cont}(h^\alpha)\; e^{-iH^\alpha t}\; F(H^\alpha<b)\; F(|y_\alpha|<R') \tag{5.41}$$

is a uniformly continuous family of compact operators. Therefore we can find $R(\varepsilon)\geq R_1$ such that for all $R>R(\varepsilon)$, $0\leq t\leq T(\varepsilon)$

$$\|F(|y_\alpha|>R)\; K(t)\| < \varepsilon\; \Big(8\; T(\varepsilon) \sum_{\beta\neq\alpha} \|V_\beta\|\Big)^{-1}. \tag{5.42}$$

This finishes the proof of Lemma 5.4. The size of $R(\varepsilon)$ and $T(\varepsilon)$ depends on ρ and the upper cutoff b of the energy support A. Otherwise it is independent of Ψ.

□

VI. Time Evolution of Certain Observables

The main difference between two body and three body scattering is that three far separated particles can have significant interaction both in the future and in the past. As an example one may think of two far separated heavy particles and a light one between them. It may happen that the light particle is scattered back and forth several times before all particles move away from each other. By energy momentum conservation the rescattering can happen only finitely many times but it is nevertheless too hard to follow the details of the time evolution. As a way out we look at certain observables which have a very peculiar feature. In the classical case their change in time is exactly the same for the free time evolution and for the time evolution of pointlike particles which interact by elastic collisions. If we divide by suitable powers of the time the effects of the interaction like time delay or the formation of resonances do not matter in the long run. Thus we will show in the quantum case that the effects of the interaction can be neglected asymptotically for the time evolution of these observables.

On the other hand we obtain enough information about the state which allows to determine a time after which the rescattering will not occur any more. Moreover we get control of the kinematic characteristics of the state: the separation of the particles, their relative kinetic energy, and their relativ direction of flight. We will study the observables in this section. The kinematical consequences will be used in the next section to show that the time evolution can be approximated by simpler ones in the far future.
Define

$$\frac{1}{2} X^2 = \frac{\mu_\alpha}{2} |x_\alpha|^2 + \frac{\nu_\alpha}{2} |y_\alpha|^2 , \tag{6.1}$$

$$D = \frac{1}{2}\{\vec{p}_\alpha \cdot \vec{x}_\alpha + \vec{x}_\alpha \cdot \vec{p}_\alpha + \vec{q}_\alpha \cdot \vec{y}_\alpha + \vec{y}_\alpha \cdot \vec{q}_\alpha\}. \tag{6.2}$$

Both operators are independent of the chosen decomposition α.
By A(t) denote the time translate of A:

$$A(t) = e^{iHt} A e^{-iHt} . \tag{6.3}$$

Proposition 6.1. On $\mathcal{H}_s$ we have in the sense of strong resolvent convergence

$$\frac{1}{2} X^2(t)/t^2 \to H, \tag{6.4}$$

$$D(t)/t \to 2H . \tag{6.5}$$

Proof. The demonstration is very similar to the two particle case [9; 7,Section VII]. We omit here the easy domain questions and we give only the essentials of the proof. A simple calculation using the canonical commutation relations shows

$$\begin{aligned} \frac{d}{dt} \frac{1}{2} X^2(t) &= e^{iHt} i[H, \tfrac{1}{2} X^2] e^{-iHt} \\ &= e^{iHt} i[H_o, \tfrac{1}{2} X^2] e^{-iHt} \\ &= D(t). \end{aligned} \tag{6.6}$$

$$\begin{aligned} \frac{d}{dt} D(t) &= e^{iHt} i[H,D] e^{-iHt} \\ &= e^{iHt}\{i[H_o,D] + i[V,D]\} e^{-iHt} \end{aligned}$$

$$= 2H_o(t) - \sum_\alpha \vec{x}_\alpha \cdot (\vec{\nabla} V_\alpha)(\vec{x}_\alpha)(t)$$

$$= 2H - \sum_\alpha I_\alpha(t) . \tag{6.7}$$

In the third step we have used for each V_α the representation of D with the same decomposition α. The commutator with the first two terms in D results in the given term whereas the other two commute with V_α. We have introduced the shorthand for the interaction term

$$I_\alpha = 2V_\alpha + \vec{x}_\alpha \cdot (\vec{\nabla} V_\alpha)(\vec{x}_\alpha) . \tag{6.8}$$

With our assumptions about the potentials (2.8 & 9) we have that the I_α are bounded operators and

$$\lim_{\rho \to \infty} \| I_\alpha F(|x_\alpha| > \rho) \| = 0 . \tag{6.9}$$

$$\frac{D(T)}{T} = \frac{D(0)}{T} + \frac{1}{T} \int_0^T dt [2H - \Sigma I_\alpha(t)]$$

$$= 2H + \frac{D(0)}{T} + \frac{1}{T} \int_0^T \sum_\alpha I_\alpha(t) dt \tag{6.10}$$

The first term is the desired limit, the second goes to zero on a suitably chosen core of H. The third term can be split using (6.9) into a term which is small in norm for all t if ρ is large enough and the remainder

$$\Sigma \frac{1}{T} \int_0^T dt \; I_\alpha(t) \; F(|x_\alpha| < \rho)(t) . \tag{6.11}$$

On a core of $H \restriction \mathcal{H}_s$ this tends to zero as $T \to \infty$ by Proposition 5.1. This implies the strong resolvent convergence (6.5). An analogous argument with a twofold time mean shows (6.4).

□

These results alone are not very useful in the case of three particle scattering. For large times according to (6.5) a state must be localized mainly in the positive spectral subspace of D. This does not yet imply, however, that the component

$$d^\alpha = \frac{1}{2}(\vec{x}_\alpha \cdot \vec{p}_\alpha + \vec{p}_\alpha \cdot \vec{x}_\alpha) \tag{6.12}$$

related to the pair alone is positive. Note that $d^\alpha > 0$ means that the particles in the pair are outgoing relative to each other. The situation is improved by the following easy observation.

<u>Theorem 6.2.</u> Let $\Psi \in \mathcal{H}_s$, $F(H \in A)\ \Psi = \Psi$, then there exists a sequence $\tau_n \to \infty$ such that far all α

$$\text{a)}\quad \| [H_0 - H]\ e^{-iH\tau_n}\ \Psi \| \to 0\ , \tag{6.13}$$

$$\text{b)}\quad \| [f(\mu_\alpha \frac{x_\alpha}{\tau_n}) - f(\vec{p}_\alpha)]\ e^{-iH\tau_n}\ \Psi \| \to 0\ , \tag{6.14}$$

$$\| [f(\nu_\alpha \frac{y_\alpha}{\tau_n}) - f(\vec{q}_\alpha)]\ e^{-iH\tau_n}\ \Psi \| \to 0\ , \tag{6.14'}$$

$$\text{c)}\quad \| [\phi\{\frac{\mu_\alpha}{2} (\frac{x_\alpha}{\tau_n})^2\} - \phi(h^\alpha)]\ e^{-iH\tau_n}\ \Psi \| \to 0\ , \tag{6.15}$$

as $n \to \infty$ for any $f \in C_0^\infty(\mathbb{R}^\nu)$ and $\phi \in C_0^\infty(\mathbb{R})$.

<u>Proof.</u> a) Let $\rho_n \to \infty$ be a sequence of increasing cutoffs, then as $n \to \infty$

$$\| [H_0 - H] \prod_\alpha F(|x_\alpha| > \rho_n) \| \to 0\ .$$

By Proposition 5.1. we can pick for each ρ_n a time τ_n such that as $n \to \infty$

$$\sum_\alpha \| F(|x_\alpha| < \rho_n)\ e^{-iH\tau_n}\ \Psi \| \to 0\ . \tag{6.16}$$

Combining the last two statements gives (6.13).

b) Let Ψ be in addition in the core of $H \upharpoonright \mathcal{H}_s$ which was used in the proof of Proposition 6.1. Then we have for the components of the vectors

$$\sum_{i=1}^{\nu} \frac{1}{2\mu_\alpha} \| (\mu_\alpha \frac{x_{\alpha,i}}{\tau_n} - p_{\alpha,i})\ e^{-iH\tau_n}\ \Psi \|^2 +$$

$$+ \sum_{i=1}^{\nu} \frac{1}{2\nu_\alpha} \| (\nu_\alpha \frac{y_{\alpha,i}}{\tau_n} - q_{\alpha,i})\ e^{-iH\tau_n}\ \Psi \|^2$$

$$= (e^{-iH\tau_n}\ \Psi,\ [\frac{1}{2} X^2 - D + H_0]\ e^{-iH\tau_n}\ \Psi)$$

$$\to (e^{-iH\tau_n}\ \Psi,\ [H - 2H + H]\ e^{-iH\tau_n}\ \Psi) = 0 \tag{6.17}$$

In particular we get for any $\vec{q} \in \mathbb{R}^\nu$ and any Ψ as specified in the theorem

$$\lim_{n\to\infty} \| (\exp\{i\vec{q}\cdot[\mu_\alpha \frac{\vec{x}_\alpha}{\tau_n} - \vec{p}_\alpha]\} - 1\!\!1)\ e^{-iH\tau_n}\ \Psi \| = 0. \tag{6.18}$$

By the Baker-Campbell-Hausdorff formula this is equivalent to

$$\lim_{n\to\infty} \| (\exp\{i\vec{q}\cdot\mu_\alpha \frac{\vec{x}_\alpha}{\tau_n}\} - \exp\{i\vec{q}\cdot\vec{p}_\alpha\})\, e^{-iH\tau_n}\Psi\| = 0. \tag{6.19}$$

Integration with the Fourier transform $\hat{f}(\vec{q})$ yields (6.14). The proof of (6.14') is similar.

c) Observe that with $h_o^\alpha = (2\mu_\alpha)^{-1}\, p_\alpha^2$

$[\phi\{\frac{\mu_\alpha}{2} (\frac{\vec{x}_\alpha}{\tau_n})^2\} - \phi(h_o^\alpha)]$ is contained in case b), and due to (6.16) we conclude that

$$\|[\phi(h_o^\alpha) - \phi(h^\alpha)]\, e^{-iH\tau_n}\Psi\| \to 0 . \tag{6.20}$$

□

This result is very strong and useful, it states that the directions of $\vec{x}_\alpha$ and $\vec{p}_\alpha$ are almost parallel in the far future and antiparallel in the remote past. In particular rescattering is no longer possible for late enough times. The conditions (6.14 & 15) determine an absorbing set for the full time evolution. On this set we can control the future time evolution. We approximate it by a simpler one in the next section.

VII. Asymptotic Evolution of Scattering States

We know that asymptotically on $\mathcal{H}_s$ the total kinetic energy is strictly positive if we assume as before $\Psi = F(H \in A)\Psi$. There are two possibilities: either all three pairs have strictly positive relative kinetic energy, or for a pair the relative energy is small, then the kinetic energy of the third particle relative to the pair is bigger. We treat the two cases by different methods. We begin with a pair having low energy. We use the smooth cutoff function ϕ which has been defined in the paragraph preceding Lemma 5.3. Then in the range of $\phi(h^\alpha)$ the pair has low energy.

Lemma 7.1. Let $\Psi \in \mathcal{H}_s$ and $F(H \in A)\,\Psi = \Psi$, the sequence τ_n as given in Theorem 6.2. Then

$$\lim_{n\to\infty} \sup_{t\geq 0} \|[e^{-iHt} - e^{-iH^\alpha t}]\, \phi^2(h^\alpha)\, e^{-iH\tau_n}\Psi\| = 0 . \tag{7.1}$$

Proof. Similar as in the beginning of the proof of Lemma 5.2 we use (6.16) to conclude that

$$\|[\phi^2(h^\alpha) - \phi^2(h^\alpha)\, g(k_o^\alpha)]\, e^{-iH\tau_n}\Psi\| \to 0 \tag{7.2}$$

as $n\to\infty$ where g is the smooth cutoff function as defined following (5.7). The functions are chosen such that g=1 on the range of $\phi^2(h^\alpha)\; F(H^\alpha \in A')$. Thus it is sufficient to consider states where the relative energy of the pair is smaller than 2a' and the kinetic energy is bigger than a-3a'. We choose again a' small enough to satisfy (5.22) with the kinematical implications and notation discussed there. By (6.14') and (5.10) the operator

$$g(k_o^\alpha) = \sum_j \chi_j^2\,(\vec{q}_\alpha) \to \sum \chi_j(\vec{q}_\alpha)\;\chi_j(\nu_\alpha\, y_\alpha/\tau_n)$$

on $\exp(-iH\tau_n)\Psi$. Therefore in (5.12) the sum over $|\vec{a}| \leq R$ does not contribute for $R = 6v_o\tau_n \leq [2(a-3a')/\nu_\alpha]^{1/2}\;\tau_n$ as $n\to\infty$. Moreover this implies that the sum (5.14) over the incoming components does not contribute asymptotically because of the alignment of y_α/τ_n with $q_\alpha\neq 0$. Thus we can replace for large n the factor $g(k_o^\alpha)$ by P_R^{out} with $R=6v_o\tau_n$. (As above P_R^{out} is used to describe the motion of the third particle relative to the pair.) Similarly by (6.14) we can replace one of the factors $\phi(h^\alpha)$ by $\phi\{\frac{\mu_\alpha}{2}\;(\frac{x_\alpha}{\tau_n})^2\}$ the range of which is contained in $|x_\alpha| \leq v_o\tau_n$. Summing up we have shown that

$$\lim_{n\to\infty} \|[\phi^2(h^\alpha) - \phi(h^\alpha)\; F(|x_\alpha| < v_o\tau_n)\; P_R^{out}]\; e^{-iH\tau_n}\,\Psi\| = 0 \qquad (7.3)$$

where $R=6v_o\tau_n$. By Lemma 5.3 we know that

$$\lim_{n\to\infty}\; \sup_{t\geq 0} \|[e^{-iHt} - e^{-iH^\alpha t}]\;\phi(h^\alpha)\; F(|x_\alpha| < v_o\tau_n)\; P_{6v_o\tau_n}^{out}\| = 0. \qquad (7.4)$$

□

The remaining term consists of that part of the state where for each pair the particles are far separated and move away from each other with strictly positive speed.

<u>Lemma 7.2.</u> With the conditions of Lemma 7.1

$$\lim_{n\to\infty}\; \sup_{t\geq 0} \|[e^{-iHt} - e^{-iH_o t}]\;\{1 - \sum_\beta \phi^2(h^\beta)\}\; e^{-iH\tau_n}\,\Psi\| = 0. \qquad (7.5)$$

<u>Proof.</u> We know from (6.20) that we can replace $\phi^2(h^\beta)$ by $\phi^2(h_o^\beta)$ and the three subtracted components are asymptotically pairwise orthogonal due to the kinematical restrictions.

Thus

$$\lim_{n\to\infty} \|[\{1 - \sum_\beta \phi^2(h^\beta)\} - \prod_\beta \bar{g}(h_o^\beta)]\; e^{-iH\tau_n}\,\Psi\| = 0, \qquad (7.6)$$

where $\bar{g} \in C_o^\infty(\mathbb{R})$ and $\bar{g}(e) = 1-\phi^2(e)$ for $e \leq b$. Note that $\bar{g}(e) = 0$ for $e \leq a'$. Similarly as in the proof of the previous Lemma it follows from (6.14) that one can replace $\bar{g}(h_o^\alpha)$ by $\bar{P}_{R,\alpha}^{out}$ with $R=\bar{v}\tau_n$. However, $\bar{P}_{R,\alpha}^{out}$ ist defined for the variables p_α and x_α analogous to (5.10 - 13) with $\bar{g}$ playing the rôle of g. Thus for a state in the range of $\bar{P}_{R,\alpha}^{out}$ the particles in the pair are outgoing relative to each other. The minimal speed is $\bar{v}=\{2a'/\max \mu_\alpha\}^{1/2}$ in this case. Thus we obtain for $R=\bar{v}\tau_n$, any α,

$$\lim_{n\to\infty} \|[\prod_\beta \bar{g}(h_o^\beta) - \bar{P}_{R,\alpha}^{out} \prod_{\beta\neq\alpha} \bar{g}(h_o^\beta)]\, e^{-iH\tau_n}\psi\| = 0. \tag{7.7}$$

The order of the factors in the subtracted expression does not matter since $\chi(\mu_\alpha \vec{x}_\alpha/\tau_n)$ and $\bar{g}(h_o^\beta)$ commute asymptotically. We use Cook's estimate to control the difference of the time evolutions in (7.5) and obtain with (7.6 & 7) as a bound

$$\int_0^\infty dt \sum_\alpha \|V_\alpha\, e^{-iH_o t}\, \bar{P}_{R,\alpha}^{out} \prod_{\beta\neq\alpha} \bar{g}(h_o^\beta)\, e^{-iH\tau_n}\psi\|$$

$$\leq \int_0^\infty dt \sum_\alpha \{\|V_\alpha\, F(|x_\alpha| > \tfrac{1}{2}(R+\bar{v}t))\| + \tag{7.8}$$

$$+ \|V_\alpha\|\cdot\|F(|x_\alpha| < \tfrac{1}{2}(R+\bar{v}t))\, e^{-ih_o^\alpha t}\, \bar{P}_{R,\alpha}^{out}\|\}$$

where $R=\bar{v}\tau_n\to\infty$ as $n\to\infty$. The first summand tends to zero by the decay assumption (2.8) on the potentials. The rapid decay of the second integrand in $R+\bar{v}t$ is a well known fact for the free time evolution (e.g. Lemma 5.3 in [8]). This completes the proof of Lemma 7.2.

□

Now we are prepared to prove our main result.

<u>Proof. of Theorem 3.1.</u> We verify (3.3) and (3.4). For given $\Phi \in \mathcal{H}^{cont}$ decompose it orthogonally

$$\Phi = \sum_\alpha \Psi^\alpha + \Psi', \quad \Psi_\alpha \in \operatorname{Ran} \Omega_-^\alpha, \quad \Psi' \in \mathcal{H}_s .$$

Then we know that

$$\lim_{\tau\to\infty} \sup_{t\geq 0} \|(e^{-iHt} - e^{-iH^\alpha t})\, e^{-iH\tau}\, \Psi^\alpha\| = 0 .$$

For any $\varepsilon > 0$ determine parameters $0<a<b<\infty$ and the corresponding energy support A such that for $\Psi=F(H\in A)\Psi'$ we have $\|\Psi-\Psi'\| < \varepsilon$. Decompose Ψ according to Lemmas 7.1 und 7.2 and pick a time τ such that

$$\Phi^\alpha(\tau) = e^{-iH\tau}\, \Psi^\alpha + \phi^2(h^\alpha)\, e^{-iH\tau}\, \Psi$$

satisfies (3.4) and similarly for

$$\Phi^{o}(\tau) = [1 - \phi^{2}(h^{\alpha})]\, e^{-iH\tau}\, \Psi \; .$$

We have given the proof for the future. The analogous result for the past follows by interchanging the rôle of incoming and outgoing parts.

□

References

1. W.O. Amrein and V. Georgescu, Helv. Phys. Acta 46, 635-658(1973).

2. W.O. Amrein, D.B. Pearson, M. Wollenberg, Helv. Phys. Acta 53, 335-351(1980).

3. E.B. Davies, Duke Math. J. 47, 171-185(1980).

4. V. Enss, Commun. Math. Phys. 61, 285-291(1978).

5. -, Ann. Phys. (N.Y.) 119, 117-132(1979).

6. -, Commun. Math. Phys. 65, 151-165(1979).

7. -: Geometric methods in spectral and scattering theory of Schrödinger operators, in: Rigorous Atomic and Molecular Physics, G. Velo, A.S. Wightman eds., Plenum, New York 1981.

8. -, Acta Physica Austriaca, Suppl. 23, 29-63(1981).

9. -: Asymptotic observables on scattering states, preprint Dept. Math. Univ. Bochum, in preparation

10. -: Propagation properties of quantum scattering states, preprint Institut Mittag-Leffler, Djursholm, 1982

11. -: Three Body Quantum Scattering Theory, preprint Dept. Math. Univ. Bochum, in preparation.

12. L.D. Faddeev: Mathematical Aspects of the Three Body Problem in Quantum Scattering Theory, Israel Program for Scientific Translations, 1965.

13. J. Ginibre: La méthode "dépendant du temps" dans le problème de la complétude asymptotique, preprint Univ. Paris-Sud, LPTHE 80/10, 1980.

14. J. Ginibre and M. Moulin, Ann. Inst. H. Poincaré A 21, 97-145(1974).

15. S.P. Merkuriev: Acta Physica Austriaca, Suppl. 23, 65-110(1981) and references given there.

16. E. Mourre, Commun. Math. Phys. 68, 91-94(1979).

17. -, Commun. Math. Phys. 78, 391-408(1981).

18. P.A. Perry, Duke Math. J. 47, 187-193(1980).

19. P.A. Perry, I. Sigal, and B. Simon, Ann. Math. 114, 519-567(1981).

20. M. Reed and B. Simon: Methods of Modern Mathematical Physics, III. Scattering Theory, Academic Press, New York 1979.

21. -:-, IV. Analysis of Operators, Academic Press, New York 1978.

22. D. Ruelle, Nuovo Cimento 61 A, 655-662(1969).

23. B. Simon, Duke Math. J. 46, 119-168(1979).

24. D.R. Yafaev: On the proof of Enss of asymptotic completeness in potential scattering theory, to appear in Mathematics of the USSR, Sbornik.

MATHEMATICAL STRUCTURE
IN QUANTUM FIELD THEORY

John E. Roberts
Fachbereich Physik, Universität Osnabrück
Postfach 4469, D-4500 Osnabrück

1. Introduction.

This talk is primarily addressed to non-specialists and I beg the indulgence of experts in the audience for various gross oversimplifications and one-sided opinions.

I must begin by explaining how a physicist like myself should come to emphasise the mathematical structure of a physical theory rather than its predictions or its physical content.

Quantum field theory is as old as quantum mechanics but, whereas quantum mechanics reached its final form by about 1930 after a rapid burst of activity over five years, quantum field theory is still in such a state of flux that a talk of the same title given by other mathematical physicists would be likely to emphasise quite different aspects of the subject.

Of course, it is worth bearing in mind that quantum field theory is designed to cope with elementary particle physics whereas quantum mechanics was designed to cope with atomic physics so that, in retrospective, it may turn out that the slow progress in quantum field theory will be attributed not to the formidable mathematical problems that have confronted field theorists but to the need to accumulate sufficient experimental evidence before the relevant structure could be recognized. Elementary particle physics continues to evolve: many aspects have changed beyond recognition in the past ten years, others have retained their validity.

The formidable mathematical difficulties manifested themselves early, during the thirties, in the form of divergent integrals, the ultraviolet and infrared divergences. Probably, the accepted view among physicists until about 1949 was that quantum field theory was just wrong. The development of renormalized perturbation theory by Feynman, Dyson and Schwinger circumvented the divergence problem and provided a recipe for performing calculations in quantum electrodynamics. Can a theory be wrong which leads to a result for the anomalous magnetic moment for the electron $2(1+a)\,\frac{e}{2m}\,\frac{\vec{\sigma}}{2}$ where

$$a = 0.001\ 159\ 652\ 359$$

as compared with an experimental value

$$a = 0.001\ 159\ 652\ 410_{(200)}\ ?$$

To explain the formal conceptual scheme lying behind this computation needs a large textbook. Perhaps only a few percent of this scheme is in any way directly involved in the recipe.

It is only since about 1967 that there has been a serious attempt to solve the mathematical difficulties in the sense of constructing explicit non-trivial mathematical models. This programme of constructive field theory was initiated by Nelson, Glimm and Jaffe. The early, rapid and impressive progress in constructing models demonstrated clearly that the formal conceptual scheme could be realized in a non-trivial way with the expected physical properties. Unfortunately, the problem of constructing any model of direct physical interest such as quantum electrodynamics has not been solved. Whilst the situation is probably not hopeless, there are still serious problems to be overcome and even the outcome is still open: do the theories of physical interest exist as mathematical models in the sense that we believe and, if so, are we trying to construct them in the right way?

With the progress of elementary particle physics, ideas on which theories are of physical interest have begun to crystallize. Whilst the numerical success of quantum electrodynamics is still a singular event in quantum field theory, the Salam-Weinberg theory, a unified theory of weak and electromagnetic interactions, and quantum chromodynamics, a theory of strong interactions, have come to be accepted by theoretical and experimental physicists in the course of the last ten years as representing significant advances in the understanding of these fundamental interactions. Both these theories belong to a class of theories modelled on quantum electrodynamics known as gauge theories, which apparently have a definite geometric content.

A mathematician looking at the literature in field theory after the advent of gauge theories could not fail to be struck by the way that quantum field theory has apparently acquired quite new mathematical dimensions: evocative words like connection, curvature, holonomy and homotopy appear in conjunction with specific field-theoretical jargon, gauge group, confinement, Higgs mechanism or instanton. Unfortunately, there are subtle semantic problems involved when the same words are used in neighbouring disciplines. For this reason, some superficial remarks on classical field theory are unavoidable.

2. Classical Field Theory.

For our purposes, a classical field can be regarded as a section of a bundle E over space-time which can be taken to be Minkowski space $\mathbb{M}$. The bundle and hence the fields are acted on by two groups. The Poincaré group P acts on E inducing its natural affine action $x \to Lx$ on $\mathbb{M}$. Its action α on a field ϕ is thus given by

$$(\alpha_L \phi)(x) = L\phi(L^{-1}x), \qquad x \in \mathbb{M},\ L \in P.$$

The gauge group G, which could be trivial, acts as bundle automorphisms, i.e. induces the trivial action on the base $\mathbb{M}$. Its action on the fields will be denoted by β. Here are a few examples designed to act as a guide to the physicist's terminology.

α) A complex scalar field ϕ, i.e. the bundle is a complex line bundle, with $G = U(1)$ acting on ϕ by

$$(\beta_\theta \phi)\ (x) = e^{i\theta}\phi(x).$$

Here α and β commute and one speaks of a gauge group of the first kind.

β) A complex scalar field ϕ but with $G = C(\mathbb{M}, U(1))$ acting on ϕ by

$$(\beta_\theta \phi)\ (x) = e^{i\theta(x)}\phi(x).$$

Here α and β do not commute and one speaks of a gauge group of the second kind.

γ) A vector potential field A^μ with $G = C(\mathbb{M}, U(1))$ and

$$(\beta_\theta A)^\mu(x) = A^\mu(x) + \partial^\mu\theta(x).$$

If this vector potential is the only basic field one talks about a pure gauge theory to distinguish it from a more general gauge theory involving other basic fields such as the complex scalar field of example β) as well. Electrodynamics is a gauge theory where the vector potential A^μ is coupled to a spinor field ψ in place of the complex scalar field ϕ. If U(1) is replaced by a non-Abelian Lie group such as U(2) or SU(3), one speaks of a non-Abelian gauge theory or of a Yang-Mills theory. As is well known cf. [1], the language of fibre bundles is appropriate for gauge theories. Thus the gauge group G is the group of sections of a principal fibre bundle, the fields appear as sections of an associated bundle and a vector potential is a connection on the principal fibre bundle.

Of course, a classical field is not an arbitrary section of the bundle E but a section satisfying some hyperbolic partial differential equation such as

$$\Box\phi + m^2\phi + \lambda(\phi,\phi)\ \phi = 0$$

for a complex scalar field as in example α) considered here as the section of a complex Hermitian line bundle. These field equations are supposed invariant under the action of both the Poincaré group P and the gauge group G . Thus field theorists are interested not in the (trivial) bundle E as such but in the set $\Phi \subset \Gamma(E)$ of solutions of the field equations, a highly complicated infinite-dimensional object. More precisely, the intrinsic object is Φ/G rather than Φ.

To ease the transition to quantum field theory, I want you to imagine using algebraic methods to study the infinite-dimensional bundle $\Phi \to \Phi/G$. That is, Φ is replaced by an algebra $\mathfrak{F}$, the *field algebra*, of complex-valued functions on Φ. G and P will act as automorphisms of $\mathfrak{F}$. The *observable algebra* $\mathfrak{A}$ is the fixed-point algebra of $\mathfrak{F}$ under the action of G. It is an algebra of gauge-invariant functions on Φ and can thus be regarded as an algebra of functions on Φ/G. Typically, expressions such as $\phi(x)^*\phi(y)$, $\phi(x)^*\phi(x)$ or $F^{\mu\nu}(x) = \partial^\mu A^\nu(x) - \partial^\nu A^\mu(x)$ appear as elements of $\mathfrak{A}$ in examples α), β) and γ) respectively. Combining β) and γ) we have expressions such as $\phi(x)^*(\partial^\mu - iA^\mu(x))\phi(x)$ or

$$\phi(x)^* \, e^{i\int A^\mu(x')db_\mu(x')}\phi(y) \qquad (\dagger)$$

where b is a path from y to x appearing as typical elements of $\mathfrak{A}$ in an Abelian gauge theory.

Again the language of fibre bundles is appropriate for gauge theories: $F^{\mu\nu}$ is being used as the induced connection form on the associated vector bundle. Unfortunately, even when a quantum field theorist uses the word connection or one of those other evocative words I listed earlier, his understanding is strictly at this level. Their role is not even really clear at the level of Φ/G let alone in quantum field theory. Elementary particle physics points to gauge theories but geometric structure must be understood directly at the quantum level if it is to stimulate progress in quantum field theory.

3. Quantum Field Theory.

When we come to quantum field theory our systems become intrinsically non-commutative. We have to work directly with non-commutative algebras which should be thought of as the analogues of $\mathfrak{F}$ or $\mathfrak{A}$ above. Φ or Φ/G would make sense only if there is a suitable notion of spectrum for these algebras. Furthermore there is no sense in talking about $\phi(x)$ as an element of $\mathfrak{F}$ or $\phi(x)^*\phi(y)$ as an element of $\mathfrak{A}$. These objects are too singular: at best $\phi(x)$ is an unbounded operator-valued distribution, i.e. we must regularize to get unbounded operators on some Hilbert space. Even at the level of classical field theory, the question of singular solutions of non-linear classical field equations is not well posed. Hence the formidable difficulties facing constructive field theory are neither surprising nor in themselves convincing evidence for fundamental flaws in our conception of quantum field theory.

Despite our inability to exhibit the interesting quantum fields as mathematical objects, there has been significant progress in understanding the relevant mathematical structure. In the late fifties, Haag [2] made a truly remarkable break with the traditions of quantum field theory. He suggested that the basic object is the observable net $\mathfrak{A}$ assigning to a region $\mathcal{O}$ in space-time the algebra $\mathfrak{A}(\mathcal{O})$ generated by the observables which can be measured within $\mathcal{O}$. The hallmark of a quantum field theory is therefore not a field depending on points in space-time but algebras depending on regions in space-time. This is still very much of a minority view of quantum field theory but one that has stood the test of time rather well. To give some idea of the flavour of the subject, here is a fairly standard list of assumptions used in the algebraic approach to quantum field theory.

a) Net Structure. To each open set $\mathcal{O} \subset \mathbb{M}$ there is a von Neumann algebra $\mathfrak{A}(\mathcal{O})$ on a Hilbert space $\mathcal{H}_0$;

$$\mathcal{O}_1 \subset \mathcal{O}_2 \quad \text{implies} \quad \mathfrak{A}(\mathcal{O}_1) \subset \mathfrak{A}(\mathcal{O}_2).$$

b) Local Commutativity. The algebras of spacelike separated open sets commute with one another:

$$\mathcal{O}_1 \subset \mathcal{O}_2' \quad \text{implies} \quad \mathfrak{A}(\mathcal{O}_1) \subset \mathfrak{A}(\mathcal{O}_2)'.$$

c) Additivity. The von Neumann algebra of an open set is generated by the von Neumann algebras of any open cover:

$$\mathcal{O} = \bigcup_i \mathcal{O}_i \quad \text{implies} \quad \mathfrak{A}(\mathcal{O}) = \bigvee_i \mathfrak{A}(\mathcal{O}_i).$$

d) Vacuum. There is a unit vector $\Omega \in \mathcal{H}_0$ defining a pure vacuum state ω_0 on $\mathfrak{A}$:

$$\omega_0(A) = (\Omega, A\Omega), \quad A \in \mathfrak{A}.$$

e) Poincaré Covariance. There is a σ-continuous action $L \to \alpha_L$ of the Poincaré group by automorphisms of $\mathfrak{A}$ satisfying

$$\alpha_L(\mathfrak{A}(\mathcal{O})) = \mathfrak{A}(L\mathcal{O}), \quad \omega_0 \circ \alpha_L = \alpha_L, \quad L \in \mathcal{P}.$$

Hence these automorphisms are implemented by a continuous unitary representation $\mathcal{U}_0$ of $\mathcal{P}$ on $\mathcal{H}_0$:

$$\alpha_L(A) = \mathcal{U}_0(L)A\mathcal{U}_0(L)^{-1}, \quad \mathcal{U}_0(L)\Omega = \Omega.$$

f) Positivity of the Energy. The total energy-momentum 4-vector P, i.e. the generator of the restriction of $\mathcal{U}_0$ to the subgroup of space-time translations, has its spectrum in the closed forward light cone, i.e. $P^0 \geq 0$ and $P.\,P \geq 0$.

I should perhaps remark that various aspects of the assumptions, for example the existence of a vacuum state, reflect not so much the nature of a quantum field as sensible idealizations of the experimental practice of elementary particle physics.

4. Gauge Sheaf

A relatively new discovery in algebraic quantum field theory is the role played by a sheaf of von Neumann algebras. If $\mathcal{O}$ is a non-empty connected open set, define $S(\mathcal{O}) = \mathfrak{A}(\mathcal{O})'$. Then if $\mathcal{O}$ is covered by non-empty connected open sets $\mathcal{O}_i$, we have as a consequence of additivity

$$S(\mathcal{O}) = \bigcap_i S(\mathcal{O}_i), \qquad \mathcal{O} = \bigcup_i \mathcal{O}_i.$$

In fact, S becomes a sheaf of von Neumann algebras if we extend the definition to arbitrary open sets by setting $S(\emptyset) = 0$ and $S(\mathcal{O}) = \prod_i S(\mathcal{O}_i)$, where the $\mathcal{O}_i$ are the connected components of $\mathcal{O}$. S is actually a subsheaf of the constant sheaf with fibre $B(H_0)$.

S can be thought of as equipping space-time with a geometric structure determining a local geometry, a standpoint that helps to fit quantum field theory into an established pattern within mathematics. I wish to suggest here that the key to grasping this role of S lies in the concept of gauge transformation.

This concept has had a chequered history going back to H. Weyl [3] who understood it as a space-time dependent transformation of a unit of length in his unified theory of gravitation and electromagnetism. But, as Einstein pointed out, cf. [3], this theory predicts that the wave length of light in the presence of a magnetic field is path-dependent. This would lead to diffuse spectral lines contradicting experiment. A gauge transformation had to be reinterpreted as a space-dependent phase transformation of the wave function of a charged particle [4,5] before acquiring physical relevance. The resulting path-dependent phase difference in the wave function of a charged particle in the presence of a magnetic field is the basis of the Aharanov-Bohm effect [6]. Since the relevant concept of gauge transformation is quantum mechanical, it is a retrograde step to interpret it as an automorphism of a U(1)-bundle even if this generalizes naturally to non-Abelian gauge fields. Quantum field theorists do not go this far, of course, but the aesthetic appeal of the theory of fibre bundles has influenced physicists, cf. [1].

What then do quantum field theorists understand by a gauge transformation? I know of no attempt at a general definition although in the special case of a global gauge transformation of the first kind such a definition is implicit in [7]. However, any such definition would certainly require the expectation values to be invariant under gauge transformations. This principle of gauge invariance should hold for local as well as global gauge transformations. I therefore propose interpreting a unitary operator $V \in S(\mathcal{O})$ as a local gauge transformation in $\mathcal{O}$ since it transforms any wave function $\Psi \in H_0$ into a wave function $V\Psi$ equivalent for measurements within $\mathcal{O}$

$$(\Psi, A\Psi) = (V\Psi, AV\Psi), \quad A \in \mathfrak{A}(\mathcal{O}).$$

Of course, the global gauge transformations of [7] are gauge transformations in this sense in the Hilbert space of the field algebra.

This proposal has the merit of conceptual clarity in the spirit of algebraic quantum field theory and of being independent of classical field theory. Nevertheless, there is also a significant area of conflict between it and the intuitive view of a quantum analogue of bundle automorphisms. For example, local gauge transformations in the above sense are in no way peculiar to gauge theories. In fact, there appear to be no simple quantum analogues of the bundle automorphisms of a classical gauge theory even in the case of the free electromagnetic field. Attempts to define such gauge transformations rigorously testify to this [8,9].

If we set ourselves the task of 'quantizing' gauge fields defined as sections of a fibre bundle then it anyway makes good mathematical sense to demand invariance under bundle automorphisms. This alone suffices to motivate a gauge-invariant renormalization procedure, or using minimal interactions as a heuristic principle or regarding expressions such as (†) suitably regularized as good candidates for local observables in quantum field theory. Hence all these uses of 'gauge invariance' are in fact quite independent of any quantum notion of gauge transformation. On the other hand, the mere fact that (†) rather than $\phi(x)^*\phi(y)$ yields a local observable after regularization means that (†) is, by definition, gauge invariant in the sense of the above proposal and suffices to accomodate the Aharanov-Bohm effect.

By contrast, the above proposal suggests quite a different perspective: the unitary group of the gauge sheaf S is a sheaf of gauge transformations which might play a role in quantum field theory analogous to that of the sheaf of local bundle automorphisms of the trivial principal bundle in the theory of fibre bundles. Thus superselection sectors correspond to a special class of locally free Hermitian S-modules. This means that the wave functions in the various sectors arise by patching wave functions in the vacuum sector using local gauge transformations. For further details the reader is referred to [10] with the warning that the discussion of the composition of sectors in section 9 misses the basic point that the bimodule structure refers to an induced sheaf at spacelike infinity. Patching using gauge transformations plays a role in other contexts too and it is only with a quantum notion of gauge transformation that questions of gauge type in the sense of [1] or the relation between solitons and gauge transformations at spacelike infinity become well posed problems within the framework of quantum field theory.

Acknowledgements

I wish to express my sincere thanks to the Research School of Physical Sciences at the Australian National University for its financial support and to Derek Robinson for his hospitality. During my stay in Canberra my ideas on local gauge transformations matured. I have taken the liberty of including some of these ideas in the written version of this talk in place of a table of the formal structural analogy between Hermitian vector bundles and superselection sectors since it provides a conceptual basis for this analogy.

References

[1] T.T. Wu, C.N. Yang, Concept of nonintegrable phase factors and global formulation of gauge fields, Phys. Rev. D 12, 3845 - 3857 (1975).

[2] R. Haag, Discussion des "axiomes" et des propriétés asymptotiques d' une théorie des champs locale avec particules composées, Colloques Internationaux du CNRS LXXV Lille 1957, CNRS, Paris 1959.

[3] H. Weyl, Eine neue Erweiterung der Relativitätstheorie, Ann. d. Physik 59, 101 - 133 (1919).

[4] F. London, Quantenmechanische Deutung der Theorie von Weyl, Z. f. Physik 42, 375 - 389 (1927).

[5] H. Weyl, Elektron und Gravitation I, Z. f. Physik 56, 330 - 352 (1929).

[6] Y. Aharanov, D. Bohm, Significance of Electromagnetic Potentials in the Quantum Theory, Phys. Rev. 115, 485 - 491 (1959).

[7] S. Doplicher, R. Haag, J.E. Roberts, Fields, Observables and Gauge Transformations I, Commun. Math. Phys. 13, 1 - 23 (1969).

[8] F. Strocchi, A.S. Wightman, Proof of the Charge Superselection Rule in Local Relativistic Quantum Field Theory, J. Math. Phys. 15, 2198 - 2224 (1974).

[9] A.L. Carey, J.M. Gaffney, C.A. Hurst, A C*-algebra Formulation of Gauge Transformations of the Second Kind for the Electromagnetic Field, Rep. Math. Phys. 13, 419 - 436 (1978).

[10] J.E. Roberts, New Light on the Mathematical Structure of Algebraic Field Theory, Proc. Symp. Pure Math. 38(2), 523-550 (1982).

HYDRODYNAMIK GEKOPPELTER DIFFUSIONEN: FLUKTUATIONEN IM GLEICHGEWICHT

H. Rost (Heidelberg) *)

Abstract. Renormalization in space and time of a system of diffusions interacting by a pair potential at equilibrium is carried out under an additional hypothesis. The result is the identification of the transport coefficient by a simple formula in which the pair potential enters only via the first two derivatives of its partition function.

1. Einleitung

Es ist ein Gemeinplatz, daß die statistische Physik sich als ein Ziel gesetzt hat, makroskopische Gesetze für große Systeme, sei es im statischen (Gleichgewichts-) Fall, sei es im Fall eines in der Zeit ablaufenden Prozesses aus der Natur der lokalen, mikroskopischen Wechselbeziehung zwischen den einzelnen Bestandteilen des Systems herzuleiten. Die mathematischen Schwierigkeiten, welche diesem Ziel bei realen, klassischen Modellen entgegenstehen, sind wohlbekannt. Aus diesem Grund erscheint es legitim, wenn sich der Mathematiker vereinfachte, idealisierte, oder auch künstliche Modelle konstruiert, die ganz oder in Teilaspekten einer mathematischen Analyse zugänglich sind, und an welchen sich trotzdem wesentliche Züge der vermuteten physikalischen Gesetzmäßigkeit erkennen lassen.

Dieses Programm versuchen wir im vorliegenden Artikel bezogen auf das Problem der Analyse des "hydrodynamischen Limes" durchzuführen. (Wir erinnern daran, daß dieser Limes durch reine Renormierung von Raum und Zeit eines Vielteilchensystems definiert ist, im Fall harter Kugeln mit elastischen Stößen bedeutet dies, daß man die Zahl der Kugeln pro Volumeneinheit gegen unendlich streben läßt, umgekehrt proportional zur dritten Potenz des Kugelradius, sodaß stets ein nichtausgearteter Anteil des Gesamtvolumens von den Kugeln ausgefüllt ist - im Unterschied etwa zum Boltzmann-Grad-Limes konstanten freien Wegs oder zum Wlassov-Limes einer schwachen, weitreichenden Wechselwirkung.) Ein Argument für unsere Methode liegt auch in dem Umstand

*) Unterstützt von der Deutschen Forschungsgemeinschaft (SFB 123) und NATO-Grant Nr. 040.82

begründet, daß die herkömmliche Herleitung einer klassischen Euler-Gleichung, die Entwicklung nach Chapman und Enskog, zwei verschiedenartige Grenzübergänge nacheinander involviert: zuerst den Grenzübergang des verdünnten Gases, der zur Boltzmanngleichung führt und anschließend ein entgegengesetzt wirkendes Erhöhen der Kollisionshäufigkeit. Im Gegensatz dazu versuchen wir hier bei vereinfachten ("irrealen") Systemen mit einem einzigen Limes, der durch Renormierung erklärt ist, wesentliche makroskopische Eigenschaften herzuleiten.

Es gibt derzeit unseres Wissens "hydrodynamische" Beschreibungen lediglich für spezielle physikalische Systeme, die insofern ausgeartet sind, als sie unendlich viele Erhaltungsgrößen besitzen: die klassischen harten Latten in Dimension 1 ([1]) und das harmonische Kristall ([2]). In der vorliegenden Arbeit sprechen wir von einer Klasse von Prozessen mit stochastischer Evolution,bei denen nur eine Erhaltungsgröße vorhanden ist, die Teilchenzahl, von welcher man ein diffusives Verhalten nachweisen will. Es gibt zwei mögliche Richtungen der Untersuchung: eine könnte man als "Beispielsammlung" bezeichnen, als Durchrechnen konkreter Modelle auf hydrodynamisches Verhalten. Dies steht nicht im Mittelpunkt dieser Arbeit, und wir referieren ohne Beweis im Text zwei Beispiele (unter anderem). Die andere Richtung ist die Suche nach allgemeinen Relationen, welche für mögliche Grenzdynamiken nach fortgesetzter Renormierung gelten. Diesem Ziel ordnet sich das Hauptresultat der Arbeit, der Satz in § 4, unter.

Der Aufbau der Arbeit ist der folgende: in § 2 wird die betrachtete Modellklasse vorgestellt und wird der Sprachgebrauch, was wir als "hydrodynamischen Limes" verstehen wollen, eingeführt. § 3 ist auf heuristischem Niveau; er rechnet vor, wie eine makroskopische Beschreibung im Nichtgleichgewichtsfall aussehen könnte. Einfacher scheint demgegenüber die Methode von § 4 zu sein: die Analyse des Fluktuationsprozesses unter einer stationären Verteilung. Das Problem ist in diesem Fall die Bestimmung der Kovarianzstruktur zu diesen Fluktuationen nach Renormierung , oder, anders ausgedrückt, des Diffusions- oder Transportkoeffizienten für die diffundierende Masse. Das Resultat, das man als Fluktuations- Dissipationstheorem etikettieren könnte, stellt unter einer zusätzlichen Hypothese eine einfache Relation zwischen dem Transportkoeffizienten $\varkappa$, der Varianz pro Volumeneinheit der Teilchendichte ("Suszeptibilität") χ, der Teilchendichte ρ und der Varianzkonstanten des treibenden Wienerprozesses σ^2 dar:

$$\varkappa \cdot \chi = \rho \cdot \sigma^2$$

Die dazu erforderliche Hypothese kann auch für sich allein Interesse beanspruchen, sie besagt, daß Fluktuationen zufälliger lokaler Größen linear durch Fluktuationen der Teilchenzahl ausgedrückt werden können. Beweistechnisch gesehen ist ihre Rolle die, daß sie in der Semimartingaldarstellung für die entsprechenden stochastischen Felder gestattet zu schließen, daß im Limes Martingal- und Driftterm separat gegen die entsprechenden Terme des Grenzsystems konvergieren; daher erlaubt sie die quadratische Variation des Grenzprozesses zu berechnen. (Ich verdanke H. Spohn den Hinweis, daß der Anwendungsbereich der Hypothese auf bestimmte Diffusionssysteme beschränkt ist, eben solche, für die eine Itô-Gleichung der Form (21) gilt, in deren Driftterm nur zweite Ableitungen der Testfunktion stehen. Für den allgemeinen Fall siehe [8])

Ich habe vielen Personen zu danken für Anregungen und Diskussionen, aus denen die dargelegten Argumente erwachsen sind: in erster Linie Th. Brox (Heidelberg), der mich auf das Fluktuations-Dissipations-Theorem gelenkt hat, dann H.Spohn (New Brunswick), E. Presutti (Rom) und C. Boldrighini (Camerino).

§2. Problemstellung. Bezeichnungen und Definitionen

Gegeben sei die folgende Dynamik für die Entwicklung eines Vielteilchensystems in R^d, die man als die Dynamik wechselwirkender Diffusionen bezeichnen könnte :

$$(1) \qquad dX_i = c_i(X)\cdot dt + \sigma\cdot dW_i \quad , \quad i \in I ;$$

hierin bedeutet I eine endliche oder abzählbare Indexmenge,

W_i , $i \in I$, ein System unabhängiger standardisierter Wienerprozesse,

X_i die Position des Teilchens Nr. i, X die Gesamtkonfiguration,

$c_i(.)$ eine Funktion der Konfiguration, die nur vom Verhalten der Konfiguration in der Nähe von x_i abhängt; im folgenden setzen wir voraus, daß

$$(2) \qquad c_i(x) = -\frac{1}{2}\cdot \sum_{j\neq i} \nabla\Phi(x_i - x_j) \ ,$$

wo Φ symmetrisch und von endlicher Reichweite (kompaktem Träger) ist.
Wir bemerken, daß diese Dynamik invariant unter Permutation der Indices ist; somit kann man (1) als Evolution eines Punktsystems $\{X_i\}$ auffassen. Weiter ist sie verträglich mit Translationen in R^d; das unendliche System (1) besitzt für jeden Wert ρ in einem geeigneten Intervall genau einen translationsinvarianten, räumlich ergodischen Punktprozeß der Dichte ρ als reversibel-invariantes Maß. Dies wurde in [6] gezeigt, unter Glattheitsbedingungen an Φ . Wir bezeichnen dieses Maß mit μ_ρ .

Definition 1. Zu einem Prozeß der Form (1) betrachten wir das zufällige Maß N(t) in R^d, welches gegeben ist durch seine Integrale über Testfunktionen

$$(3) \quad N(t,\varphi) := \int N(t,du)\,\varphi(u) := \sum_i \varphi(X(i,t)) \quad , \quad \varphi \in \mathcal{S}(R^d) \ .$$

Wir renormieren dieses Maß in Abhängigkeit von einem Parameter ε , dem Verhältnis von mikroskopischer zu makroskopischer Längeneinheit:

Definition 2. Das zufällige Maß $N^\varepsilon(t)$ ist für $\varepsilon > 0$ gegeben durch

$$(4) \qquad N^\varepsilon(t,\varphi) := \varepsilon^d \cdot \sum_i \varphi(\varepsilon X(i,\varepsilon^{-2}\cdot t)) \quad , \quad \varphi \in \mathcal{S}.$$

Die Normierung der Zeit mit dem Faktor ε^{-2} ist motiviert durch den Umstand, daß man nur in Zeiten der Ordnung ε^{-2} nichttriviale Änderungen im makroskopischen Verhalten des Systems erwarten kann : man denke an den Fall $c_i = 0$. Andere Dynamiken, etwa vom Typ kollidierender Teilchen, die eine Geschwindigkeit besitzen, würden eine Normierung der Zeit um den Faktor ε^{-1} erfordern.

Wir wollen nun präzisieren, was wir unter einem hydrodynamischen Verhalten des Systems verstehen wollen. Als wesentlich betrachten wir die Gültigkeit von Aussagen der beiden folgenden Typen :

Satz I (Gesetz der großen Zahlen). Es gibt eine, i.a. nichtlineare, Halbgruppe mononer Abbildungen (T_t) in einem geeigneten Raum L von positiven, lokal integrablen Funktionen in R^d mit der Eigenschaft :

wenn $N^{\varepsilon}(0)$ eine Familie deterministischer Anfangsbedingungen ist, die bei $\varepsilon \to 0$ im vagen Sinn gegen das Maß $f_o(u)du$ strebt, $f_o \in L$, dann strebt für alle $t > 0$ und $\varphi \in \mathcal{S}$ die Zufallsvariable $N^{\varepsilon}(t,\varphi)$ stochastisch gegen $\int f(t,u)\varphi(u)du$. Hierbei ist $f(t,.) = T_t f_o$.

Wenn die hier erwähnte Halbgruppe (T_t) einen Generator A besitzt, so nennt man die Evolutionsgleichung für $f(t,.)$

(5) $$\frac{\partial}{\partial t} f = Af$$

die <u>kinetische Gleichung</u> für das makroskopische Verhalten des Systems. Man könnte sie auch als "Euler-Gleichung" bezeichnen; doch erscheint es vielleicht ratsam, diesen Terminus zu vermeiden, um Verwechslungen mit Begriffen aus der "realen" Physik vorzubeugen. Typischerweise wird man bei Modellen des Typs (1) eine kinetische Gleichung mit rechter Seite

(6) $$Af = \frac{1}{2} \cdot \nabla(\varkappa(f) \cdot \nabla f)$$

erwarten. In diesem Fall könnte man $\varkappa(\rho)$ als den dichteabhängigen <u>Diffusionskoeffizienten</u> bezeichnen.

Die Aussage des nächsten Typs geht über das mikroskopische Verhalten des Systems in der Nähe eines beliebigen makroskopischen Orts u ; dazu definiert man

<u>Definition 3</u>. Für $t > 0$, $u \in R^d$ und $\varepsilon > 0$ ist $N^{\varepsilon}(t,u)$ das zufällige Maß, dessen Integrale gegeben sind durch

(7) $$N^{\varepsilon}(t,u;\varphi) = \sum_i \varphi(X(i,\varepsilon^{-2} \cdot t) - \varepsilon^{-1} \cdot u) , \quad \varphi \in \mathcal{S} .$$

<u>Satz II</u> (Lokales Gleichgewicht). Unter den Voraussetzungen von Satz I strebt die Verteilung des Punktprozesses $N^{\varepsilon}(t,u)$ für alle $t > 0$, $u \in R^d$ schwach gegen das Gleichgewichtsmaß $\mu_{f(t,u)}$. ($f(t,u)$ wie in Satz I).

Als <u>Beispiele</u> für hydrodynamisches Verhalten können bis jetzt erst wenige Modelle dienen, die exakt durchgerechnet worden sind. Wir erwähnen hier

1) ein diskretes Analogon, die <u>symmetrische Irrfahrt mit einfachem Ausschluß</u> (simple exclusion random walk) : die Teilchen bewegen sich in Z^d; jedes von ihnen springt nach einer exponentiellen Zeit mit Parameter λ , und zwar mit gleicher Wahrscheinlichkeit auf eines der 2d Nachbarfelder; ist dieses Feld besetzt, so unterbleibt der Sprung. Die Gleichgewichtsmaße, welche räumlich ergodisch sind, sind hier die <u>Bernoullimaße</u> (definiert dadurch, daß die Gitterplätze unabhängig mit gleicher Wahrscheinlichkeit mit einem Teilchen besetzt sind). Die kinetische Gleichung lautet
$$\frac{\partial}{\partial t} f = \frac{\lambda}{2} \cdot \Delta f.$$
Sie ist linear, da die Einerkorrelationen in diesem einfachen Modell einer geschlossenen Gleichung genügen. (Siehe [5], [7])

2) den Prozeß der eindimensionalen diffundierenden harten Latten (hard rods):

$$dX_i = \sigma \cdot dW_i \quad , \ i \in I \ ,$$

mit elastischer Reflexion am Rand des Phasenraums, welcher durch die Bedingung $|x_i - x_j| > c$ für alle $i \neq j$ gekennzeichnet ist. Die kinetische Gleichung ist hier von der Fron (6) mit

$$\varkappa(\rho) = \sigma^2 \cdot (1 - c\rho)^{-2} \ ; \tag{8}$$

das Gleichgewichtsmaß μ_ρ ist der Erneuerungsprozeß in R , dessen Abstände gleich c plus eine Exponentialvariable des Parameters λ sind, $\rho = (c + \lambda^{-1})^{-1}$.
(Unveröffentlichtes Manuskript des Verfassers).

§3. Ein heuristisches Prinzip zur Herleitung der kinetischen Gleichung.

In diesem Abschnitt versuchen wir für die Dynamik der Form (1) und (2) das makroskopische Verhalten zu bestimmen; es wird vorausgesetzt, daß man bereits a priori weiß, daß ein Gesetz der großen Zahl und die Eigenschaft des lokalen Gleichgewichts gelten. Wir führen der Einfachheit halber die Betrachtung nur für d=1 durch.
Die Idee besteht darin, die zeitliche Entwicklung der Zufallsvariablen $N(t,\varphi)$ in der Form einer Itô-Gleichung zu schreiben (die zugrundeliegende Filtration ist die von den Prozessen W_i erzeugte) :

$$dN(t,\varphi) = dt \cdot \left\{ \sum_i c_i(X) \cdot \nabla\varphi(X_i) + \sum_i \frac{\sigma^2}{2} \cdot \Delta\varphi(X_i) \right\} + \sigma \cdot \sum_i \nabla\varphi(X_i) \cdot dW_i \ . \tag{9}$$

Unter Benutzung von (2) behandeln wir den ersten Summanden rechts weiter; er ist

$$-\frac{1}{2} \sum_i \sum_{j \neq i} \nabla\varphi(X_i) \cdot \nabla\Phi(X_i - X_j) = \frac{1}{2} \cdot \sum_{\{i,j\}} \nabla\Phi(X_i - X_j) \cdot (\nabla\varphi(X_j) - \nabla\varphi(X_i))$$

(Summation über alle ungeordneten Paare i,j)

$$= \frac{1}{2} \cdot \sum_{\{i,j\}} \nabla\Phi(X_i - X_j) \cdot \int_{X_i}^{X_j} \Delta\varphi(u)\,du = \frac{1}{2} \cdot \langle \omega(X), \Delta\varphi \rangle \ .$$

Hierin ist $\omega(X)$ ein, von der Konfiguration X abhängiges, daher zufälliges, Maß in R, welches schreibbar ist als

$$\omega(x) = \sum \omega_{ij}(x) \quad \text{mit}$$

$$\omega_{ij}(x,du) = \begin{cases} \Phi(x_i - x_j) \cdot 1_{\{x_i < u < x_j\}} du \ , & \text{falls } x_i < x_j \\ \Phi(x_j - x_i) \cdot 1_{\{x_j < u < x_i\}} du & \text{sonst.} \end{cases}$$

Das Maß ω zählt Eigenschaften von Paaren der Konfiguration X; es ist räumlich stationär, falls der Punktprozeß $\{X_i\}$ es ist. Bezeichnen wir mit $a(\rho)$ seinen Mittelwert pro Volumeneinheit unter der Gleichgewichtsverteilung μ_ρ für $\{X_i\}$.

Wir renormieren nunmehr (9) und erhalten eine Itô-Gleichung für $N^\varepsilon(t,\varphi)$:

$$dN^\varepsilon(t,\varphi) = \frac{1}{2} \cdot dt \left\{ \langle \omega^\varepsilon, \Delta\varphi \rangle + \sigma^2 \cdot N^\varepsilon(t,\Delta\varphi) \right\} + \varepsilon^{d} \cdot \sum \nabla\varphi(\varepsilon X_i) \cdot dW_i^\varepsilon \ , \tag{10}$$

wo $W_i^\varepsilon(.)$ wieder eine standardisierte Brownsche Bewegung ist, nämlich $\varepsilon \cdot W_i(\varepsilon^{-2} \cdot \ .)$,

$$\langle \omega^{\varepsilon}, \psi \rangle = \varepsilon^{d} \cdot \langle \omega, \psi(\varepsilon .) \rangle .$$

Falls ein Gesetz der großen Zahlen gilt, geht der Martingalbestandteil in (10) nach null; seine quadratische Variation pro Zeit wird dann nämlich asymptotisch gleich

$$\varepsilon^{d} \cdot \sigma^{2} \cdot \int (\nabla \varphi(u))^{2} \cdot f(t,u)\,du \quad ;$$

für den Driftterm in (10) gilt, unter der Hypothese des lokalen Gleichgewichts, zusätzlich zu einer Annahme einer lokalen Unabhängigkeit, daß er approximativ gleich

$$(11) \qquad \frac{1}{2} \cdot \int \left[a(f(t,u)) + \sigma^{2} \cdot f(t,u) \right] \cdot \Delta\varphi(u)\,du$$

ist. Dies führt auf eine Gleichung

$$(12) \qquad \frac{d}{dt}\left(\int f(t,u)\,\varphi(u)\,du \right) = \frac{1}{2} \cdot \int (a(f(t,u)) + \sigma^{2} \cdot f(t,u)) \cdot \Delta\varphi(u)\,du ,$$

welches die schwache Form von

$$(13) \qquad \frac{\partial}{\partial t} f = \frac{1}{2} \cdot \nabla (\varkappa(f) \cdot \nabla f)$$

ist, sofern man setzt

$$(14) \qquad \varkappa(\rho) = \sigma^{2} + \frac{da}{d\rho} .$$

Diese Überlegungen lassen sich natürlich auch in höherer Dimension anstellen ; die rechte Seite von (12) bekommt dann die Gestalt

$$(12a) \qquad \frac{1}{2} \cdot \int \sum_{k,l=1}^{d} (a_{kl}(f(t,u) + \sigma^{2} \cdot \delta_{kl} \cdot f(t,u)) \cdot \varphi_{kl}(u)\,du,$$

(φ_{kl} : zweite gemischte Ableitung von φ), wo aber keineswegs klar ist, daß der Tensor a_{kl} isotrop , d.h. gleich $a \cdot \delta_{kl}$, ist, sodaß man nicht sofort auf eine kinetische Gleichung vom Typ (6) schließen kann.

§4. Bestimmung des Diffusionskoeffizienten aus dem Prozeß der Fluktuationen im Gleichgewicht.

Fixieren wir einen zulässigen Wert ρ der Dichte des Teilchensystems (1) im Gleichgewicht. Wir betrachten nun den zugehörigen stationären Prozeß mit der Verteilung μ_{ρ} zu jeder festen Zeit; alle Mittelwerte verstehen sich im folgenden als bezüglich dieses Prozesses gebildet, bzw. bezüglich μ_{ρ} , falls es sich um statische Größen handelt.

Wir führen nunmehr die Fluktuationen der Teilchenzahl um ihren Mittelwert in der Normierung des zentralen Grenzwertsatzes ein; es sind dies Prozesse $\hat{N}^{\varepsilon}$ mit Werten im Raum der Distributionen in R^{d} ; ihre Auswertung an der Stelle $\varphi \in \mathcal{S}$ ist definiert als

<u>Definition 4.</u> $\hat{N}^{\varepsilon}(t,\varphi) := \varepsilon^{d/2} \left\{ \sum_{i} \varphi(\varepsilon X(i,\varepsilon^{-2}t) - \varepsilon^{-d} \cdot \rho \cdot \int \varphi(u)\,du \right\}$, $t \geq 0$, $\varphi \in \mathcal{S}$.

Für jedes t ist $\hat{N}^{\varepsilon}(t,\varphi)$ zentriert und von nicht ausgearteter Varianz bei $\varepsilon \to 0$.
Im Fall eines glatten Φ mit endlicher Reichweite gilt ein zentraler Grenzwertsatz der Gleichgewichtsmaße ([3]) ; der Limes ist trivial, nämlich ein <u>weißes Rauschen</u>, d.h. <u>für festes t</u> ist $\hat{N}^{\varepsilon}(t,\varphi)$ asymptotisch Gaußisch mit Varianz

$$(15) \qquad \mathcal{E}(\hat{N}(t,\varphi))^{2} = \chi \cdot \int \varphi^{2}(u)\,du .$$

(Wir erinnern daran, daß das Gleichgewichtsmaß μ_ρ Gibbsmaß ist zum Paarpotential Φ mit der Temperaturkonstanten $\beta = \sigma^{-2}$, zu einem geeigneten chemischen Potential α, d.h. formal geschrieben werden kann als (s [6])

$$(16) \qquad \mu_\rho(dx) = c\cdot dx\cdot \exp(-\sigma^{-2}\cdot \sum_{\{i,j\}} \Phi(x_i - x_j) + \alpha\cdot \sum_i 1) .$$

Wenn $\Psi(\alpha)$ den zugehörigen normierten Logarithmus der Zustandssumme, den "Druck", bezeichnet, so hat man die Beziehung

$$(17) \qquad \rho = \frac{d\Psi}{d\alpha} , \qquad \chi = \left(\frac{d}{d\alpha}\right)^2 \Psi ,$$

welche α und damit die Varianzkonstante χ in (15) zu gegebenem ρ festlegt.)

Das Ziel ist es nun, die Kovarianz des Prozesses $\hat{N}^\varepsilon$ <u>über verschiedene Zeiten</u> zu identifizieren, im Limes $\varepsilon \to 0$. Es wird nämlich allgemein, aus "physikalischen" Erwägungen, geglaubt, daß der Diffusionskoeffizient $\varkappa(\rho)$ sich aus dem Limes der Kovarianz von $\hat{N}^\varepsilon$ (im Gleichgewicht μ_ρ) bestimmen läßt durch

$$(18) \qquad \lim_\varepsilon \mathcal{E}\, \hat{N}^\varepsilon(0,\varphi)\cdot \hat{N}^\varepsilon(t,\psi) = \chi\cdot \iint \varphi(x)\psi(y) g(\varkappa(\rho)\cdot t, x-y)\, dxdy$$

($g(a,.)$: Gaußische Dichte der Varianz a, zentriert) .

Diese Identifikation möglicher Limiten (der Kovarianzen) von $\hat{N}^\varepsilon$ ist allerdings nur unter einer zusätzlichen Annahme möglich, die in hinreichender Allgemeinheit für Prozesse der Form (1) und (2) zu beweisen zur Zeit nicht möglich erscheint. Um diese Annahme einzuführen, müssen wir zunächst einige Umformungen vornehmen. Wir schreiben die Dynamik von (15) in Itô-scher Form, zunächst für $d = 1$:

$$(19) \qquad d\hat{N}^\varepsilon(t,\varphi) = \frac{1}{2}\cdot dt\cdot \left[\langle \omega^\varepsilon, \Delta\varphi\rangle\, \varepsilon^{-d/2} + \sigma^2\cdot \hat{N}^\varepsilon(t,\Delta\varphi)\right] + \varepsilon^{d/2}\cdot \sum \nabla\varphi(\varepsilon X_i)\cdot \sigma\cdot dW_i^\varepsilon .$$

Wir führen auch für die Maße ω bzw. ω^ε die Normierung des zentralen Grenzwertsatzes ein und zentrieren sie :

$$(20) \qquad \hat{\omega}^\varepsilon(du) := \varepsilon^{-d/2}(\omega^\varepsilon(du) - a(\rho)du) .$$

Da das Integral von $\Delta\varphi$ über R^d verschwindet, geht (19) über in

$$(21) \qquad d\hat{N}^\varepsilon(t,\varphi) = \frac{1}{2}\cdot dt\cdot \left[\langle \hat{\omega}^\varepsilon, \Delta\varphi\rangle + \sigma^2\cdot \hat{N}^\varepsilon(t,\Delta\varphi)\right] + \varepsilon^{d/2}\cdot \sum \nabla\varphi(\varepsilon X_i)\cdot \sigma\cdot dW_i^\varepsilon$$

Wenn man vom Fall $d = 1$ abweicht, geht der Driftterm über in

$$(21a) \qquad \frac{1}{2}\cdot dt\cdot \left[\sum_{k,l} \langle \hat{\omega}^\varepsilon_{kl}, \varphi_{kl}\rangle + \sigma^2\cdot \delta_{kl}\cdot \hat{N}^\varepsilon(t,\varphi_{kl})\right] ,$$

wo die $\hat{\omega}^\varepsilon_{kl}$ analog zu (20) gebildet sind.

Die nun folgende Hypothese spiegelt den aus der Nichtgleichgewichtstheorie vertrauten Sachverhalt wider, daß alle lokalen Mittelwerte durch den Wert der Erhaltungsgrößen an den jeweiligen Stellen bestimmt sind - und zwar modifiziert für den Fall der <u>zufälligen</u> Fluktuationen um die Gleichgewichtslage. Sie ist verifiziert für einfach durchzurechnende Modelle (z.B. [9]) und dürfte relativ allgemein gelten.

<u>Hypothese</u>. Es gibt Zahlen γ_{kl}, $k,l = 1,\dots,d$, derart daß für alle $\varphi \in \mathcal{S}$ die Variablen $\langle \hat{\omega}^\varepsilon_{kl}(t), \varphi\rangle$ in folgendem Sinn durch $\gamma_{kl}\cdot \hat{N}^\varepsilon(t,\varphi)$ ersetzt werden können :

(22) für alle $t>0$ strebt bei $\varepsilon \to 0$ der Ausdruck

$$\int_0^t (\langle \hat{\omega}^{\varepsilon}_{kl}(s),\varphi\rangle - \gamma_{kl}\cdot \hat{N}^{\varepsilon}(s,\varphi))ds$$ stochastisch gegen null.

Bemerkung. Da die zufälligen Felder $\hat{\omega}^{\varepsilon}_{kl}$ auch im Limes $\varepsilon \to 0$ vom Feld $\hat{N}^{\varepsilon}$ verschieden sind, kann man nicht erwarten, daß der Integrand in (22) für jedes einzelne s gegen null geht. Die hier gewählte Form der Hypothese ist aber durchaus plausibel : sie gründet sich auf die Anschauung, daß der Bestandteil von $\hat{\omega}^{\varepsilon}$, der nicht durch die "Erhaltungsgröße" $\hat{N}^{\varepsilon}$ ausdrückbar ist, schnell oszilliert : jedenfalls so schnell daß zufällige Abweichungen vom Mittelwert in einer Zeit, die wesentlich kürzer ist alsdieEinheit der makroskopischen Zeit, relaxieren.

Satz. (Fluktuations-Dissipations-Theorem)
Unter der Hypothese (H) gilt für jeden möglichen Grenzprozeß (im Sinn der schwachen Konvergenz für distributionenwertige Prozesse) $\hat{N}$ von $\hat{N}^{\varepsilon}$:

(23) $$\mathcal{E}\,\hat{N}(\varphi,0)\cdot\hat{N}(\psi,t) = \chi\cdot \iint \varphi(x)\psi(y)g(\varkappa\cdot t,x-y)dxdy$$

für alle $t > 0$, φ, $\psi \in \mathcal{S}$, wobei

(24) $$\varkappa = \rho\cdot\sigma^2\cdot\chi^{-1} .$$

Beweis. Für jedes φ ist der Prozeß

(25) $$\hat{N}^{\varepsilon}(t,\varphi) - \frac{1}{2}\cdot \int_0^t \sum_{k,l} (\langle\hat{\omega}^{\varepsilon}_{kl}(s),\varphi\rangle + \sigma^2\cdot\delta_{kl}\cdot\hat{N}^{\varepsilon}(s,\varphi_{kl}))ds$$

ein stetiges Martingal, bezüglich der von $\hat{N}^{\varepsilon}$ erzeugten Filtration, mit einer quadratischen Variation pro Zeit gleich

$$\sigma^2\cdot\varepsilon^d\cdot \sum_i |\nabla\varphi(\varepsilon X_i)|^2 .$$

Dieser Ausdruck strebt wegen der räumlichen Ergodizität von μ_ρ für jede feste Zeit nach

$$\sigma^2\cdot\rho\cdot \int |\nabla\varphi(u)|^2 du ;$$

somit und vermöge der Hypothese ist unter jedem möglichen Limesprozeß für alle $\varphi\in\mathcal{S}$ der Prozeß

(26) $$\hat{N}(t,\varphi) - \frac{1}{2}\cdot\int_0^t \sum_{k,l} (\gamma_{kl} + \sigma^2\cdot\delta_{kl})\cdot\hat{N}(s,\varphi_{kl})ds$$

ein stetiges Martingal, dessen quadratische Variation pro Zeit konstant ist, und zwar gleich

(27) $$\rho\cdot\sigma^2\cdot \int|\nabla\varphi(u)|^2 du = -\rho\cdot\sigma^2\cdot\int \Delta\varphi\cdot\varphi\, du .$$

Durch (26) und (27) wird aber ein Ornstein-Uhlenbeck-Prozeß (siehe [4]) beschrieben, welcher ebenso wie alle $\hat{N}^{\varepsilon}$ stationär ist. Sein invariantes Maß ist der Limes der invarianten Maße der $\hat{N}^{\varepsilon}$, nämlich das durch (15) gegebene weiße Rauschen. Nun gilt aber folgende einfache algebraische Identität :

wenn $\hat{N}$ so beschaffen ist, daß

$$\hat{N}(t,\varphi) - \int_0^t \hat{N}(s,A\varphi)ds$$

für alle $\varphi \in \mathcal{S}$ ein stetiges Martingal ist mit quadratischer Variation

$$\int \varphi(u) Q\varphi(u) du \ ,$$

und wenn $\hat{N}$ ein stationäres Maß hat mit Kovarianz

$$\mathcal{E}(\hat{N}(t,\varphi)^2) = \int \varphi(u) C\varphi(u) du \ ,$$

wo Q und C positive Operatoren in $\mathcal{S}$ sind, A beliebig, so gilt

(28) $$A^* C + CA = -Q \ .$$

(Dies ist einfach zu sehen : man schreibe eine Itô-Gleichung für $\hat{N}(t,\varphi)^2$ und bedenke, daß unter dem stationären Maß die Erwartung der Drift verschwindet.) In unserem Fall ergibt diese Identität, wegen

$$Q = -\rho \cdot \sigma^2 \cdot \Delta \quad \text{und} \quad C = \chi \cdot 1 \quad , \text{daß} \quad A = \frac{1}{2} \cdot \varkappa \cdot \Delta \quad {}^{*)}$$

sein muß, mit $\varkappa$ wie in (24) behauptet.

Mehr noch, wegen der Eindeutigkeit des Ornstein-Uhlenbeck-Prozesses zu diesem A und Q folgt daraus weiter, daß auch die Kovarianz des Limesprozesses $\hat{N}$ die im Ornstein-Uhlenbeck-Prozeß berechnete sein muß, wie in (23) behauptet wurde. Man hat damit auch - unter der Hypothese- den gemeinsamen Gaußischen Charakter der Variablen $\hat{N}^\varepsilon(t,\varphi)$ im Limes, für verschiedene t, bewiesen.

Bemerkungen.

1) An dieser Bestimmung des Transportkoeffizienten $\varkappa$ ist auffallend, daß er einzig und allein durch die beiden ersten Ableitungen des Drucks Ψ bestimmt ist, d.h. daß in das makroskopische Verhalten des Systems (1) die Hamiltonfunktion Φ nur durch das thermodynamische Potential eingeht, welches zur Paarwechselwirkung $\sigma^{-2} \cdot \Phi$ gebildet wird. Insbesondere verhält sich das renormierte System isotrop, obwohl von Φ lediglich Spiegelungssymmetrie vorausgesetzt war. Entscheidend ist hierfür die Isotropie des treibenden Wienerprozesses.

2) Die Heuristik des §3 paßt gut mit den Gleichgewichtsbetrachtungen dieses Abschnitts zusammen, wie man sich im Fall d=1 folgendermaßen klarmachen kann : die einzig plausible Konstante γ welche der Hypothese genügen kann, ist $\gamma = \frac{da}{d\rho}$; die Zahl γ hat nämlich anzugeben, wie man das zufällige Maß $\omega(V)$ eines großen Volumens V in affiner Weise durch die Teilchenzahl in diesem Volumen schätzen kann; nun beschreibt aber $\frac{da}{d\rho}$ die mittlere Zunahme von $\omega(V)/V$ bezogen auf die mittlere Zunahme der Teilchenzahl pro Volumen, jedenfalls in der Nähe von ρ . Diese Überlegung führt, in Übereinstimmung mit (14) auf

$$\varkappa = \gamma + \sigma^2 = \frac{da}{d\rho} + \sigma^2 \ .$$

*) A ist symmetrisch, da alle Prozesse $\hat{N}^\varepsilon$ reversibel sind.

Literaturverzeichnis

[1] C.Boldrighini, R.L.Dobrushin, Yu.A.Sukhov : The one-dimensional hard rod caricature on hydrodynamics. Preprint 1982.

[2] C.Boldrighini, A.Pellegrinotti, L.Triolo : Hydrodynamics of the one-dimensional harmonic chain. Preprint 1981.

[3] Th. Brox : Grenzverteilungen für Gibbsisch erzeugte Punktprozesse mit schwacher Wechselwirkung. Diplomarbeit. Heidelberg 1977.

[4] R.Holley, D.Stroock : Generalized Ornstein-Uhlenbeck processes and infinite particle branching Brownian motions. Publications Res. Inst. Math. Sciences 14, 741-788 (1978).

[5] A.Galves, C.Kipnis, C.Marchioro, E.Presutti : Nonequilibrium measures which exhibit a temperature gradient : study of a model. Commun. math. Physics 81, 127-147 (1981).

[6] R. Lang : Unendlichdimensionale Wienerprozesse in Wechselwirkung I.II. Z Wahrscheinlichkeitstheorie verw. Geb. 38, 55-72 (1977); 39, 277-299(1977).

[7] F.Spitzer : Recurrent random walk of an infinite particle system. Transactions Am. math. Soc. 198, 191-199 (1974).

[8] H.Spohn : Large scale behavior of equilibrium time correlation functions for some stochastic Ising models. Preprint 1982.

[9] A.Galves, C.Kipnis, H.Spohn : Fluctuation theory for the symmetric simple exclusion process. Preprint 1982.

SUPERRENORMALIZABLE INFRA RED THEORIES

R. SENEOR

Centre de Physique Théorique de l'Ecole Polytechnique
Plateau de Palaiseau - 91128 Palaiseau - Cedex - France

"Groupe de Recherche du C.N.R.S. N° 48"

INTRODUCTION

The last ten years were characterized by great successes in the rigourous construction of ultraviolet superrenormalizable field theories. Various methods ([G-J_1], [F-O], [M-S_1], [B], [G-J-S]) more or less related to the study of the thermodynamic limit in statistical mechanics were developped showing the existence of the theory in the range of the parameters corresponding to high or low temperature regions (in the equivalent statistical mechanics language). Another of the main feature of all these methodes was to strongly incorporate the inputs given by perturbation theory. In particular, the finite number of divergent diagrams was a fundamental ingredient.

At this stage of progress, attempts were done towards strictly renormalizable theories, precisely Φ^4 in 4 dimensions (cf. [G-J_2] , [S]) and towards the behaviour at the critical points (massless field theories) ([G-J_2]) but were not very successful. Corresponding to critical theories a typical problem is the behaviour of a $\lambda\Phi^4$ field theory in d dimensions, $2 \leqslant d$, for λ taking some critical values insuring a long range behaviour. For d=2 it is expected that there is a unique value λ_c (with a fixed bare mass) such that the two point correlation function has a power law decrease with exponent 1/4 (in universality with the Ising model from which the field theory can be obtained by a limiting procedure). The behaviour of the two-point function becomes only canonical, i.e. its decrease with exponent d-2 for dimensions $d > 4$, is the appearance of an anomalous dimension η correcting the exponent of decrease as $d-2+\eta$. The two region $d > 4$ and $d < 4$ can be understood on the level of perturbation theory. One can introduce an infra-red power (I-R) counting for Φ^4 diagrams with propagator $\frac{1}{p^2}$. The infra-red degree of divergence D_I at a vertex is given by $D_I = -D_{U.V} = -(d-4)$. The theory is then said infra-red superrenormalizable for $d > 4$. For $d < 4$ the strength of the I-R divergences is too strong meaning that an expansion with usual free propagator i.e. around the usual gaussian measure (in the standard functionnal language), is akward. We thus need to introduce a modified free measure linked to the anomalous dimension. However such a program involves many new tools and I only will present here some new methods allowing to deal the case of superrenormalizable I-R interactions hoping they could be useful in the general treatment of I-R theories. This work was done in collaboration with J. Magnen.

THE MODEL

Instead of Φ^4 in $d > 4$ we deal with a $|\vec{\nabla}\Phi|^4$ and work in Euclidean space $d \gtrsim 3$ in the functionnal integration formalism introduced in constructive field theory (see [E]). The free measure is the usual massless one with the Fourier transform of the covariance given by $\frac{1}{p^2}$. The gradients acting on Φ in the interaction make that on the contrary of Φ^4 in $d > 4$ there is no need of a mass quadratic term to annihilate the dynamical mass generated by the interaction. Introducing vector fields $A_\alpha(x) = \vec{\nabla}_\alpha \Phi(x) \equiv \frac{d}{dx_\alpha} \Phi(x)$ we are lead to a vectorial A^4 theory with a free measure μ of covariance $\tilde{C}_{\alpha\beta}(p) = \frac{p_\alpha p_\beta}{p^2}$. Moreover the theory is non renormalizable on the ultra violet (U.V) side and we introduce a momentum cut off (not shown explicitly here). The lattice (as an U.V. cutoff) version of this model has already being studied by many people ([B-F-L-S],[F_1], [F_2], [G-K]). The more systematic study is the one of [G-K] which make to work the renormalization group program (R-G) in the frame work introduced by [B].

As results we show, see [M-S_2], that the thermodynamic limit of this theory exists and behaves for weak coupling as a perturbed standard free measure, this measure being finitly renormalized. Consequently the large distance behaviour of the 2 point function is given by

$$\frac{1}{4\pi}\,(1+0(\lambda))\,\frac{1}{|x-y|^{d-2}} + 0\left(\frac{1}{|x-y|^{d}}\right)$$

where $1+0(\lambda)$ is a finite field strength renormalization and $0(\lambda) \to 0$ when $\lambda \to 0$.

THE PRINCIPLES

The proof is based on an expansion. The momenta are divided in ranges. In each slice of momentum range a cluster expansion in cubes scaled to the size of the range gives the thermodynamic limit. The coupling between different ranges of the momenta is controlled by a truncated perturbation expansion which takes account of the superrenormalizable aspect of the interaction. This last point is quite similar to what happen in the U.V. problem : it is a phase space expansion analogous in principle with the block spin analysis of the R.G. technics. By their own nature each elementary step of the expansion has to produce small factors. These factors have to be good enough in order to ensure the convergence of the expansion. Two features introduce difficulties in the proof of the convergence.

The first one is intrinsic to the theory. In fact it comes from the generation by perturbation of terms proportionnal to $|\vec{A}|^2$. There are two possible attitudes with respect to them : or we introduce them in the free measure modifying slightly the coefficient in front of the covariance, or we can introduce quadratic counterterms in the interaction which will cancel these terms.

The second one has a more technical origin as explained now. Our method is based on the progressive introduction of low momenta. Each time a new momentum range is introduced one makes a cluster expansion scaled in appropriate way with this range. It may happen that besides 2 point functions, there are 4 point functions which are produced with four low momentum external lines and high momenta internal ones. They behave like $\lambda|A|^4$ terms with no other small coefficients. If too many such terms are produced the gaussian contraction of them will not be controllable (this divergence is the one of the perturbation series of a Φ^4 theory ([J], [C-R])). To circumvent this difficulty one has to use the positivity of the interaction allowing to dominate the external fields by

$$\left(\frac{1}{|\Delta|}\int_\Delta A(x)\,dx\right)^n \exp - \int_\Delta |A|^4(x)\,dx \leqslant |\Delta|^{-\frac{n}{4}}\, n! \tag{1}$$

where Δ is some cube which side of size $|\Delta|^{1/3}$ is roughly the inverse of the momentum range of $A(x)$. The main object of a cluster expansion is to exhibit explicitly the control of the translation invariance of clusters containing fixed (original) fields. On the level of elementary steps it reduces in our case in having uniform bounds on translations in scaled cubes. Here it means essentially that we have to sum over all possible external legs in cubes Δ. We see by (1) and under the hypothesis of a strong enough connectedness of the n-point function that for $n > 4$ we have a factor smaller than $|\Delta|^{-1}$ which is exactly what bounds uniformly the integration of a smooth function in a cube of size $[|\Delta|^{1/3}]^3 = |\Delta|$. For n = 4, we get exactly $|\Delta|^{-1}$ without any extra factors which contradicts the principle of a convergent expansion and it follows that we need to renormalize the 4 point functions to eliminate these terms. The case n = 2 is as expected also divergent and need, as previously explained, to be renormalized. The finite renormalization counter terms correspond to the difference between bare and effective coupling and field strength. Let us remark that the fact there is no coupling constant renormalization in the U.V. problem, for example Φ_3^4, can be understood as follows. By momentum conservation a 4-point function with external legs at momentum M_i cannot have all its internal momenta lower than M_i, one has to be higher than M_i. From the superrenormalizability, this higher momentum produce a convergent factor which will improve the convergence of the external legs. This is not the case for the I-R superrenormalizability since, M_i being small, the internal momenta can be of order unity.

SOME TECHNICAL POINTS

With $\vec{A}(x) = \vec{\nabla}\Phi(x)$ one wants to compute

$$\lim_{\Lambda\to\mathbb{R}^3} \langle\Phi(x_1)\ldots\Phi(x_n)\rangle^{(\Lambda)}_{\lambda,\alpha,\delta}$$

where the expectations are taken with respect to the measure

$$d\mu^{(\Lambda)}_{\lambda,\alpha,\delta} = [Z^{(\Lambda)}_{\lambda,\alpha,\delta}]^{-1} \exp - \{(\lambda+\delta)\int_\Lambda |A|^4(x) - \frac{\alpha}{2}\int_\Lambda |A|^2(x)\}\, d\mu(\Phi) \tag{2}$$

$d\mu$ being the Gaussian measure of mean 0 and the Fourier transform of the covariance $\tilde{C}(p) = \frac{1}{p^2}$, and $Z^{(\Lambda)}_{\lambda,\alpha,\delta}$ being the normalization.*

Limiting our attention to the 2 point function (n = 2) and after integration by part

$$\langle\Phi(x)\;\Phi(y)\rangle^{(\Lambda)}_{\lambda,\alpha,\delta} = C(x-y) + \sum_{\alpha,\beta=0}^{2} \int du\, dv \frac{d}{du_\alpha} C(x-u) \frac{d}{dv_\beta} C(v-y) \left\langle\frac{\delta}{\delta A_\alpha(u)} \frac{\delta}{\delta A_\beta(v)}\right\rangle^{(\Lambda)}_{\lambda,\alpha,\delta} \tag{3}$$

The expansion is defined to control the second term. The expectation value is totally expressed in term of the $\vec{A}$'s by replacing in (2) the Gaussian measure $d\mu(\Phi)$ by $d\nu(A)$ with the Fourier transform of the covariance given by $\frac{p_\alpha p_\beta}{p^2}$.

---

* We do not specify the U.V. cutoff.

A sequence of momentum range $i \to [M_{i+1}, M_i]$, $M_i > M_{i+1}$ $i = 0,1,2,\dots,\rho$ is introduced, $M_{\rho+1}$ being on I-R cutoff to be removed at the end of the expansion and $M_o = \infty$. To this decomposition is associated a decomposition of the covariance and of the field $A = \bigoplus_i A_i$. The expansion is essentially a sequence of cluster expansions done in each momentum range in appropriate lattices (for the range i the lattice is of size $|\Delta|^{1/3} \simeq M_i^{-1/3}$). The coupling of the different momenta (through the interaction A^4) is controlled by truncated perturbation expansions. These expansions generate 2 and 4 point functions which need to be renormalized. The normalization constants are computed recursively. For example, for momentum between M_o and M_i, the constant α_i is computed (by an explicit fixed point theorem) in such a way that an amputated 2 point functions $\langle \frac{\delta}{\delta A(x)} \frac{\delta}{\delta A(y)} \rangle$ has a Fourier transform $\tilde{G}(p)$ vanishing at $p=0$. From $\tilde{G}(p) = \tilde{G}(p) - \tilde{G}(0) = p^2 F(p)$ by parity and with $F(0)$ bounded, one sees that the expectation in the r.h.s. of (3) produces 2 more derivations which after integration by part act on the covariances. It then follows using the decrease $|\vec{\nabla}^n C(x)| \sim \frac{1}{|x|^{n+1}}$, that the large distance behaviour of the 2 point function is given by $C(x-y)$ i.e. is $\frac{1}{|x-y|}$.

All these remarks are developped in [M-S_2]. Notice that by rescaling the fields in (2) one can start with a measure s.t. $\alpha=\delta=0$ and the theory are developped allows to compute wave and coupling function renormalization constants (in the multiplicative meaning).

CONCLUSION

Our method which can be thought as a R.G. for pedestrians applies to Φ^4 in more than 4 dimensions, to the dipole gas and probably to all I-R superrenormalizable situations.

REFERENCES

[B] Benfatto, G., Cassandro, M., Gallavotti, G., Nicolo, F., Oliveri, E., Pressutti, E., Scacciatelli, E., Comm. Math. Phys. 59, 143-166 (1980).

[B-F-L-S] Bricmont, J., Fontaine, J.R., Lebowitz, J., Spencer, T.,
I. Commun. Math. Phys. 78, 281 (1980)
II. Commun. Math. Phys. 78, 363 (1981).

[C-R] De Calan, C., Rivasseau, V., Comm. Math. Phys. 83, 77-82 (1982).

[E] Lecture Notes in Physics, Vol. 25, Springer (1973).

[F_1] Federbush, P., Commun. Math. Phys. 81, 327, 341 (1981).

[F_2] Fontaine, J.R., Bounds on the decay of correlations for $\lambda(\nabla\Phi)^4$ models Université catholique de Louvain (Preprint).

[F-O] Feldman, J., Osterwalder, K., Ann. Phys. 97, 80-135 (1976).

[G-J_1] Glimm, J., Jaffe, A., Fort. der Physik 21, 327 (1973).

[G-J_2] Glimm, J., Jaffe, A., Quantum Physics, Springer Verlag, Berlin-Heidelberg-New York (1981).

[G-J-S] Glimm, J., Jaffe, A., Spencer, T., in Lecture Notes in Physics, Vol. 25, Springer (1973).

[G-K] Gadwedzki, K., Kupiainen, A., Commun. Math. Phys. 82, 407 (1981).
Commun. Math. Phys. 83, 469 (1982).

[J] Jaffe, A., Commun. Math. Phys. 1, 127-149 (1965).

[M-S_1] Magnen, J., Seneor, R., Ann. Inst. H. Poincaré 24, 95 (1976).

[M-S_2] Magnen, J., Seneor, R., The Infra-red behaviour of $(\nabla\phi)^4_3$, Preprint Ecole Polytechnique, Juin 1982.

[S] Schrader, R., Commun. Math. Phys. 49, 131-153 ; 50, 97-102 (1976) ; Ann. Inst. H. Poincaré 26, 295-301.

UN MODELE D'UNIVERS CONFRONTE AUX OBSERVATIONS

Jean-Marie Souriau

Université de Provence et Centre de Physique Théorique, Marseille

---

Introduction

Le modèle d'Univers que nous présentons ici a été élaboré en plusieurs stades:

1°) Nous avons étudié la signification physique et géométrique de la CONSTANTE COSMOLOGIQUE (Souriau 1964, 1974, 1976);

2°) Un travail de colorimétrie, de photométrie et d'analyse statistique (Fliche-Souriau 1979), nous a montré qu'il n'est pas nécessaire de supposer une évolution statistique du phénomène "quasar" pour interpréter les données optiques relatives à ces objets, à condition d'adopter UN CERTAIN MODELE DE FRIEDMANN - LEMAITRE; il s'agit d'un modèle à courbure spatiale positive et à expansion indéfinie ayant commencé par le Big-Bang (ces caractéristiques sont compatibles si on ne choisit pas la valeur 0 pour la constante cosmologique).

3°) Dans les modèles de ce type, l'espace est une hypersphère S3 . Cette hypothèse permet de renouveler la "cosmologie symétrique": on peut envisager que matière et antimatière soient réparties sur deux hémisphères de S3 . La zone de contact doit alors être observable comme zone d'absence dans la répartition spatiale des quasars. NOUS AVONS EFFECTIVEMENT MIS EN EVIDENCE UNE TELLE ZONE D'ABSENCE (Souriau 1980, Fliche-Souriau-Triay 1982 I et II).

4°) Un Univers de ce type pourrait présenter une ANISOTROPIE GENERALE: nous avons effectivement trouvé divers indices d'une telle anisotropie, en particulier dans la répartition, l'orientation et la cinématique des galaxies.

Le présent travail est une étude détaillée de ce modèle et de certaines de ses conséquences. Nous commençons (§1) par la description de la zone d'absence, et montrons qu'elle suffit à déterminer un modèle cosmologique. Chacun des paragraphes suivants (§§ 2 à 17) est consacré à un type distinct d'observations, qui est confronté avec ce modèle: âge et densité de l'Univers, relation redshift - luminosité pour les galaxies et pour les quasars, relation redshift - diamètre pour les radio-sources, isotropie du rayonnement à 3°K , physique de la zone de contact matière - antimatière (largeur théorique de cette zone, taux et spectre des rayons gamma produits). Nous étudions particulièrement une éventuelle stratification de l'Univers parallèlement à cette zone, interprétant dans ce sens les raies d'absorption dans les spectres des quasars, la disposition du Super - Amas Local et du Groupe Local de galaxies, l'orientation des régions H I des galaxies, et enfin la cinématique des galaxies proches.

Pour alléger l'exposé principal, nous avons renvoyé aux annexes la plus grande partie du travail technique.

UN MODELE D'UNIVERS CONFRONTE AUX OBSERVATIONS

------§1------

La REPARTITION SPATIALE DES QUASARS, telle que nous la connaissons aujourd'hui, comporte une ZONE D'ABSENCE, que nous désignerons par (μ).

(μ) a la forme d'une coquille, dont l'épaisseur est de de 120 Mpc environ; notre regard la traverse dans une moitié du ciel; le point le plus proche est dans la direction

(α) : (17h 45m 30s, -6° 50') (1950)

à 2700 Mpc de nous (les longueurs sont évaluées ici avec H_0 = 100 Km/s/Mpc); la distance de (μ) croît régulièrement quand la visée s'éloigne de (α) .

Pour restituer la répartition SPATIALE des quasars à partir des observations (ascension droite, déclinaison, redshift), il faut avoir CHOISI UN MODELE D'UNIVERS.

Si on prend le modèle élémentaire de Hubble (espace euclidien, distance proportionnelle au redshift z), la zone d'absence (μ) apparaît comme approximativement sphérique, et dirigeant sa concavité vers nous.

Mais les grands redshifts des objets qui bordent (μ) (jusqu'à z > 2.6) requièrent l'emploi d'un modèle plus précis. Nous choisirons les modèles de FRIEDMANN-LEMAITRE, parce que ce sont les seuls modèles relativistes qui soient compatibles avec les propriétés observées du rayonnement cosmologique.

Ces modèles dépendent de DEUX PARAMETRES, dont il va falloir déterminer la valeur: le paramètre de densité Ω_0 (sans dimensions) et la CONSTANTE COSMOLOGIQUE Λ (nous ne comptons pas ici le paramètre d'échelle, la constante de Hubble H_0 , dont la détermination est indépendante).

Il existe une tradition bien établie qui affirme que Λ doit être nulle, mais avec des justifications assez faibles: l'argument d'autorité (Einstein a dit que...; tout le monde fait comme ça...); des arguments de type "simplicité", "grands nombres" (si on les poussait un peu loin, ils montreraient vite que la constante de Newton, G , est nulle elle aussi...). Peut-être est-ce simplement pour éviter des difficultés techniques qu'on préfère $\Lambda = 0$? Nous allons voir pourtant que ces difficultés ne sont pas considérables.

En fait, le principe de relativité générale, et plus généralement la théorie des groupes, ne donnent aucune raison de supposer $\Lambda = 0$, choix qui ne correspond à aucune symétrie particulière; l'hygiène épistémologique recommande donc d'introduire Λ dans le problème, et de voir si les observations permettent de la mesurer.

Pour cette mesure, les propriétés de la zone (μ) apportent une possibilité nouvelle. En effet, les données disponibles sur les quasars montrent que les deux hypothèses suivantes sont compatibles:

a) L'univers peut se décrire par un modèle de Friedmann-Lemaître;
b) La zone d'absence (μ) est plane;

et, mieux, que ces deux conditions DETERMINENT le modèle (à l'échelle près): ceci se vérifie simplement par un double balayage sur les paramètres à déterminer.

On peut donc mesurer la constante cosmologique et le paramètre de densité en utilisant comme seule donnée un catalogue de quasars; voici le résultat:

$$\Omega_\circ \# 0.1, \quad \Lambda \# (1600 \text{ Mpc})^{-2};$$

il est indiqué d'utiliser, au lieu de Λ, le PARAMETRE DE DECELERATION $q_\circ$, qui est défini avec plus de précision:

$$q_\circ = -1.12 \pm 0.01$$

Bien entendu cette approche soulève quelques questions essentielles:

- la zone (μ) survivra-t-elle aux observations à venir?
- quelle est l'interprétation physique de cette zone?
- que signifie la condition b) ci-dessus?
- quelles sont les implications du modèle d'Univers ainsi déterminé?
- est-il en accord avec les autres observations?

nous allons examiner les réponses qu'on peut proposer aujourd'hui.

Dans ce modèle, la COURBURE DE L'ESPACE est POSITIVE, et l'espace lui-même a la forme d'une HYPERSPHERE S3 (attention! avec une dimension de moins, l'analogue serait la SURFACE d'une sphère ordinaire S2 , et PAS SON INTERIEUR. Il est commode de "plonger" S3 dans un espace numérique à quatre dimensions, mais il s'agit d'un artifice mathématique, et les points situés en dehors de S3 sont fictifs).

Comme il se doit (ci-dessus b)), la zone d'absence (μ) est bien un plan, mais un plan tel qu'on peut le définir en géométrie RIEMANNIENNE; on peut se le représenter comme un EQUATEUR de S3 (analogie: toute ligne tracée sur la Terre qui apparaît comme "droite" aux observateurs proches est en fait un équateur, équateur que l'on peut évidemment associer à deux pôles diamétralement opposés).

Cette surface équatoriale (μ) sépare l'espace en deux régions - deux HEMISPHERES de S3 ; nous-mêmes, nous occupons une position assez quelconque dans l'un de ces hémisphères, avec une latitude de 27 degrés environ. La distance maximum à laquelle nous observons des quasars est à peu près de 90 degrés autour de S3 , et par conséquent les quasars s'étendent largement au delà de la zone (μ) , ce qui permet de la détecter.

Le modèle permet d'évaluer la QUANTITE TOTALE DE MATIERE existant dans l'univers: approximativement 1.2 E 80 atomes d'hydrogène, soit 1 E 23 masses solaires.

Le calcul de l'EVOLUTION TEMPORELLE montre qu'il s'agit d'un modèle à BIG-BANG (l'extrapolation vers le passé est limitée par une singularité où la température est infinie et où l'espace est réduit à un point); l'expansion est destinée à durer éternellement, avec une légère accélération (le paramètre de décélération $q_\circ$ est négatif).

Nous allons maintenant confronter ce modèle aux observations disponibles.

UN MODELE D'UNIVERS CONFRONTE AUX OBSERVATIONS

------§2------

Dans ce modèle (1), l'AGE DE L'UNIVERS (la durée écoulée depuis le Big-Bang) est de 16 MILLIARDS D'ANNEES (avec H_0 = 100); cette valeur est compatible avec les âges estimés des étoiles et des amas globulaires.

------§3------

La valeur 0.1 du PARAMETRE DE DENSITE Ω_0 obtenue dans (1) est non seulement positive (ce qui n'avait rien d'évident a priori si l'on se souvient de la méthode de détermination du modèle), mais aussi en accord avec les estimations de densité obtenues directement par l'étude des galaxies.

------§4------

Le modèle (1) est compatible avec la relation REDSHIFT-LUMINOSITE observée pour les GALAXIES.

------§5------

La relation REDSHIFT - LUMINOSITE des quasars, interprétée dans ce modèle, signifie que ces objets ont commencé à apparaître quand l'univers était âgé de 2.5 milliards d'années, et qu'ils constituent depuis un PHENOMENE STATIONNAIRE, en ce sens que leur nombre et la répartition de leurs luminosités ne dépendent pas significativement de la date; en particulier, la luminosité intrinsèque des quasars les plus brillants est remarquablement indépendante du redshift.

Ce comportement est celui qu'on peut prévoir si les quasars sont des phénomènes explosifs, peut-être de durée relativement courte et répétitifs, qui se produisent dans des noyaux de galaxies. Dans ce cas les premiers quasars n'ont pu apparaître qu'après le délai nécessaire à la constitution de ces galaxies; or 2.5 milliards d'années semble une durée raisonnable pour cette constitution.

On peut penser que le phénomène "quasar" relève d'un processus physique relativement standard; ceci parce que les spectres des quasars se ressemblent beaucoup, indépendamment du redshift auquel ils sont observés. Si c'est vrai, le nombre et l'intensité des quasars n'ont pas de raison d'avoir subi d'évolution statistique prépondérante à partir de la date de leur apparition - et c'est bien ce qu'on constate en utilisant le modèle (1) .

Au contraire, avec les modèles à constante cosmologique nulle, on est obligé d'admettre une double évolution de la statistique des quasars (évolution en nombre et en luminosité), dont l'interprétation est problématique.

------§6------

Un autre test cosmologique classique est fourni par la relation REDSHIFT -

DIAMETRE pour les radio-sources. Bien que ce test ne soit pas très précis, on constate qu'il donne des résultats au moins aussi satisfaisants avec ce modèle qu'avec ses concurrents.

------§7------

Dans les modèles à big-bang, on admet l'existence d'un stade où l'Univers était rempli d'hydrogène assez chaud pour être ionisé et par conséquent fortement couplé avec le rayonnement thermique; ce plasma était donc OPAQUE (action des ions sur le rayonnement) et VISQUEUX (action du rayonnement thermique sur les ions).

L'expansion produit un refroidissement perpétuel de l'Univers (à cause du travail fourni par la pression de radiation pour gonfler l'espace); lorsque la température est tombée en dessous d'un seuil suffisant (3000°K), l'ionisation de l'hydrogène a cessé, donc aussi sa viscosité et son opacité. Double conséquence: la matière a pu commencer à se condenser gravitationnellement, et le rayonnement libéré s'est propagé en tous sens. Il constitue le RAYONNEMENT COSMOLOGIQUE, refroidi aujourd'hui à 2.7°K .

L'observation de ce rayonnement montre qu'il est REMARQUABLEMENT ISOTROPE (ses propriétés sont les mêmes dans toutes les directions, à une petite correction près qui s'interprète par l'effet Doppler-Fizeau correspondant à un mouvement propre de la Terre).

Les sources de ce rayonnement sont extrêmement lointaines et primitives (il s'agit d'hydrogène au redshift $z = 1100$, mais tout de même pas du big-bang, comme on l'affirme parfois); par conséquent l'interprétation de leur isotropie (ou d'une légère anisotropie si on en découvre une) pose un important problème d'interprétation.

Avec un modèle à courbure NEGATIVE OU NULLE, l'espace et la matière constituant l'Univers sont censés apparaître simultanément dans un volume INFINI. La partie de cet Univers qui nous est accessible par l'observation augmente constamment, mais son volume reste fini, donc négligeable devant l'infinité inconnaissable; et pourtant le modèle postule la symétrie parfaite de cet infini. Un tel modèle n'est donc qu'une extrapolation schématique de nos observations, acceptable seulement à titre provisoire. Il rend compte de la symétrie du rayonnement (nous voyons l'Univers symétrique parce qu'il est symétrique...), mais ne l'explique pas; en effet les sources de ce rayonnement que l'on observe dans les diverses directions du ciel sont à des distances mutuelles bien trop grandes (par rapport à l'âge correspondant de l'Univers) pour que leur homogénéité puisse être la conséquence d'un processus causal.

Par contre, dans un modèle à COURBURE POSITIVE, la quantité de matière existant dans l'Univers est FINIE, et a occupé au stade initial un volume très petit (Lemaître parle en ce sens de "l'atome primitif"). On peut donc supposer que l'apparition de la matière ne pose pas de problème de causalité, même si l'extrapolation du modèle vers les conditions "initiales" ne fournit pas de réponse à ce problème. Il est clair que cette extrapolation doit être menée avec prudence - ne serait-ce que parce que le modèle ne connait pas d'autre pression que celle des photons et qu'il néglige toute production d'entropie (voir ci-dessous (1*10)).

UN MODELE D'UNIVERS CONFRONTE AUX OBSERVATIONS

Avec le modèle (1) ci-dessus, il existe une raison supplémentaire pour l'isotropie du rayonnement; le calcul montre en effet que la distance parcourue par la lumière depuis l'époque du découplage se trouve aujourd'hui voisine d'un demi-tour autour de l'espace S3 , donc que le rayonnement cosmologique observé dans toutes les directions du ciel est originaire D'UNE SEULE PETITE REGION DE L'ESPACE, proche de nos "antipodes" cosmiques; aucun problème causal n'est donc posé par l'isotropie de ces sources.

------§8------

Une observation fondamentale en cosmologie est la NEUTRALITE ELECTRIQUE de l'univers (elle est exacte à 1 E -40 près, sinon la répulsion électro-statique rendrait imperceptible l'attraction gravitationnelle).

La physique nous enseigne qu'une période initiale très chaude implique la présence, à côté de la matière, d'une grande quantité d'ANTIMATIERE - il suffit que l'énergie des collisions thermiques soit suffisante pour produire le phénomène de CREATION DE PAIRES (proton-antiproton, électron-positon, etc.). Ce processus respecte rigoureusement la neutralité électrique.

Ainsi, dans le modèle classique du big-bang, l'univers est passé par un état peuplé d'un mélange matière - antimatière (baryons - antibaryons), avec une densité nucléaire, une température de 3 E 12 °K , et une pression valant 2 E 20 fois celle qui règne au centre du Soleil.

Dans la seconde qui a suivi, trois ou quatre choses essentielles se sont produites:

- A cause de cette pression, l'expansion a été explosive, et a entraîné une réfrigération à 1 E 10 °K;

- pression et densité ont diminué d'un facteur 1 E 10;

- à cause du refroidissement, la quasi-totalité de la matière s'est annihilée par recombinaison avec l'antimatière.

Nous observons aujourd'hui un Univers électriquement neutre, et nous constatons dans notre environnement la présence de matière sans antimatière. Comment expliquer l'état actuel à partir du stade décrit plus haut? On peut proposer trois réponses principales:

- a) On peut supposer que la symétrie entre matière et antimatière n'est pas totale, donc que c'est par suite d'une loi physique que seule la matière a survécu dans tout l'espace. Cette loi pourrait être obtenue par la "Grande Unification", théorie en voie de développement dans laquelle la conservation de la matière peut être violée. Une difficulté de cette approche est de comprendre pourquoi la conservation de l'électricité, justement, n'est pas violée.

- b) On peut rester dans le cadre de la physique actuelle - et supposer l'existence initiale d'un excès de baryons et d'électrons, dosé pour être rigoureusement neutre. Après la disparition de toute l'antimatière, cet excès s'est retrouvé présent, et constitue la matière actuelle. Il s'agit d'un scénario cohérent, mais encore une fois de type NON EXPLICATIF (: "la matière est là parce qu'elle a toujours été là, et elle est neutre parce qu'elle a toujours été neutre...").

- c) On peut enfin envisager que toute la matière existante est issue de création de paires, mais que la recombinaison n'a pas été totale par suite d'irrégularités de répartition; il doit donc subsister de la matière ou de l'antimatière suivant les régions. Un milliardième environ de la matière aurait ainsi survécu - et autant d'antimatière ailleurs.

Une particularité intéressante de cette hypothèse est d'interpréter la neutralité électrique. Remarquons la différence entre cette neutralité électrique (qui s'observe partout) et la neutralité baryonique (qui ne se manifeste que par un bilan global); cette différence résulte évidemment de la portée infinie des forces électromagnétiques - alors que les autres interactions n'ont lieu qu'au contact direct.

C'est cette hypothèse qu'on appelle la "cosmogonie symétrique", et que nous allons envisager.

Dans le cas d'un univers à courbure négative ou nulle, les dissymétries initiales que l'on peut imaginer sont de type turbulent, ce qui conduit à une répartition matière - antimatière de type "émulsion". L'évolution d'une telle émulsion a été étudiée en détails (R. Omnès 1979): les surfaces de contact matière - antimatière sont le siège de réactions de recombinaison, qui ont pour effet de diminuer la courbure de ces surfaces et de faire croître la dimension des cellules . Ce processus, appelé COALESCENCE, aurait pu conduire à l'état actuel. Mais la confrontation avec les observations soulève un certain nombre de difficultés (Voir par exemple Steigman 1979).

Prenons maintenant le cas du modèle (1) ci-dessus. Le volume de l'Univers étant fini, on peut envisager l'existence d'irrégularités initiales MACROSCOPIQUES. Le cas le plus simple est une anisotropie initiale principalement DIPOLAIRE: imaginons l'évolution correspondante.

A la fin de la période de recombinaison (t = quelques secondes) matière et antimatière sont complètement séparées, et se trouvent réparties principalement en deux régions distinctes de S3. Les régions plus petites qui ont pu se former sont rapidement mangées par les deux grandes - qui viennent au contact. Le processus de coalescence régularise leur frontière commune; cette coalescence prend fin lorsque matière et antimatière sont réparties entre deux hémisphères de S3 , et en contact sur un équateur (les deux régions ne peuvent évidemment plus croître, et la courbure de la zone de contact est nulle). Alors la dématérialisation sur cet équateur adopte un régime beaucoup plus calme (voir le § 9).

Dans ce cas on s'attend donc à l'existence d'une zone équatoriale singulière - QUI SERAIT L'INTERPRETATION DE LA ZONE (μ) QUE NOUS AVONS DETECTEE. Nous allons donc étudier les conséquences observables d'un scénario de ce type.

Première remarque: la séparation matière - antimatière est ici acquise dans un stade très primitif (dès la fin de la recombinaison, donc au bout des premières secondes); elle doit donc être suivie immédiatement de la SYNTHESE DES ELEMENTS LEGERS (deutérium, hélium, lithium, etc.) - qui se termine au bout de quelques dizaines de minutes, selon un processus bien étudié. Ce processus est impliqué ici comme dans la cosmogonie classique (ci-dessus b)); on sait qu'il conduit à des abondances des éléments légers assez conformes aux observations faites dans notre environnement. Cette concordance, en cours de vérification, constitue actuellement LE SEUL TEMOIN DIRECT d'une période initiale très chaude de l'Univers.

UN MODELE D'UNIVERS CONFRONTE AUX OBSERVATIONS

------§9------

Etudions de plus près, dans ce scénario, ce qui se passe près du plan de contact matière - antimatière. Cette zone est le siège permanent de réactions de recombinaison proton - antiproton, qui produisent essentiellement des rayons gamma; ces gammas, à leur tour, interagissent avec la matière avoisinante et l'ionisent.

Le découplage matière - rayonnement, qui s'est produit à la fin de l'ionisation, a donc été RETARDE au voisinage de ce plan - jusqu'à ce que la pression et la température aient suffisamment diminué. Puisque ce découplage est préalable à toute condensation, il est possible que l'occasion de former des galaxies ait été définitivement perdue; pas de quasars, a fortiori, au voisinage.

Ce processus a été proposé par Evry Schatzman (1982), qui en a construit un modèle détaillé. Ce modèle permet de calculer en particulier la LARGEUR de la zone dépourvue de galaxies: LE RESULTAT (105 Mpc) COINCIDE TRES PRECISEMENT AVEC LA LARGEUR OBSERVEE DE LA ZONE (μ) - (un peu moins de 120 Mpc); à priori, il aurait pu exister un désaccord de plusieurs ordres de grandeur.

------§10------

L'Univers est actuellement transparent pour les rayons gamma produits dans la zone de contact matière-antimatière; on peut donc évaluer théoriquement l'INTENSITE et le SPECTRE du rayonnement que nous devons en recevoir. Schatzman (1982) a montré que ce rayonnement doit être sensiblement isotrope, par suite d'un effet de compensation (les rayons qui ont perdu le plus d'énergie par suite du redshift ont été produits aux dates les plus anciennes, donc avec le taux le plus élevé), et a évalué son flux.

Or nous disposons d'une observation dans ce domaine: le satellite SAS2 a détecté un fond continu de rayons gamma, approximativement isotrope, dont le spectre correspond à une production par dématérialisation proton-antiproton (Fichtel et al. 1978). SON TAUX EST EN ACCORD AVEC LA PREVISION THEORIQUE DE SCHATZMAN - alors que la théorie de l'émulsion conduisait à un flux de gammas très supérieur à celui qui est observé.

Deux tests observationnels positifs ((9) et (10)) sont donc en accord avec l'interprétation matière-antimatière formulée au §8.

------§11------

De simples considérations de symétrie montrent que l'existence d'une zone d'absence plane (μ) peut être associée à une STRATIFICATION GENERALE de l'Univers, parallèle à cette zone. La REPARTITION SPATIALE DES Q.S.O. semble manifester une telle stratification; toutefois une plus grande précision dans la détermination des redshifts de ces objets, techniquement possible, semble nécessaire pour confirmer cet effet.

------§12------

Les RAIES D'ABSORPTION LARGES observées dans les spectres de quelques quasars peuvent s'interpréter par la présence de nuages interposés ayant une profondeur cosmologique. En particulier, quatre de ces nuages semblent situés DANS LA ZONE (μ) ; ce sont d'ailleurs eux qui nous ont permis de la localiser initialement; ils manifestent donc, dans cette zone, un début de condensation n'ayant pas atteint le stade quasar. Quant aux autres nuages de ce type, leur répartition spatiale semble corrélée avec la stratification des Q.S.O.

------§13------

Vue de la Terre, cette stratification doit apparaître comme une ANISOTROPIE du ciel - autour du point (α) ci-dessus et de son antipode (ω) (qui est proche de Bételgeuse).

Or LES GALAXIES PROCHES NE SONT PAS EQUIPARTIES DANS LE CIEL, MAIS CONCENTREES SUR UNE ZONE ASSEZ ETROITE, axée justement sur cette direction (α) - (ω) ; en particulier (α) est proche du pôle supergalactique qui a été donné par De Vaucouleurs et al. (1976).

Cette simple constatation suggère que la stratification de l'Univers se manifeste encore à l'échelle de la répartition des galaxies voisines.

Une étude plus précise de notre environnement fait apparaître une stratification à plusieurs niveaux: le Super-Amas Local, les strates intérieures à ce super-amas mises en évidence par Tully (1982), le Groupe Local de galaxies, sont des structures aplaties et parallèles; leur direction commune ne diffère pas significativement de celle que nous venons d'indiquer.

------§14------

Passons à une échelle encore plus petite: on peut se demander si les plans des GALAXIES SPIRALES sont réellement orientés au hasard, ou préférentiellement dans telle ou telle direction.

L'étude d'un échantillon par Mac Gillivray et al.(1982) conclut effectivement à une corrélation avec la direction du Super-Amas, mais cette corrélation est à la limite de la significativité.

Il faut évidemment tenir compte ici du GAUCHISSEMENT des galaxies spirales: très fréquemment les orbites circulaires de la matière dans une galaxie ne sont pas coplanaires, et la direction de leur plan évolue à mesure que l'on s'éloigne du centre. Ce phénomène est parfois dû à l'interaction avec une galaxie proche, mais il apparaît souvent en l'absence de tout objet visible, et avec une grande symétrie par rapport au centre (voir par exemple l'étude de M33 par Sandage et al. (1980)).

Le gauchissement constitue donc un raccordement entre deux directions de plan, caractéristiques l'une de la région centrale, l'autre de la région externe.

Examinons si ce ne sont pas ces régions externes qui seraient le mieux corrélées à la stratification ambiante (§13).

Ces régions sont difficiles à observer optiquement; cependant, dans le cas des galaxies les plus proches, on y observe un certain nombre de galaxies naines "satellites". Mais la radio-astronomie fournit plus d'informations: on détecte en effet des nuages H I , constituant ce qu'on appelle le HALO de la galaxie. L'interaction de la région centrale avec le halo pourrait d'ailleurs expliquer l'existence et la permanence des bras spiraux (Ostriker et Peebles 1973).

Nous allons donc tester l'hypothèse suivante: les halos sont des structures APLATIES PARALLELEMENT à LA STRATIFICATION GENERALE.

Cette hypothèse ne pourra se vérifier que pour les galaxies que nous observons dans une direction à peu près parallèle au plan de stratification: elle implique en effet que le halo sera alors vu par la tranche, donc allongé en perspective sur le plan du ciel.

Mais il se trouve que cette circonstance favorable est TRES FREQUEMMENT REALISEE; ceci résulte simplement de la répartition dans le ciel des galaxies proches, telle que nous l'avons décrite au § 13. Ainsi 90 % des halos choisis et étudiés par Rots (1980) comme par Bosma (1981) sont situés dans la moitié du ciel la plus favorable à ce test. L'hypothèse est donc vérifiable statistiquement; et effectivement L'ORIENTATION DANS LE CIEL DE CES HALOS EST TRES FORTEMENT CORRELEE AVEC LA DIRECTION DE LA ZONE (μ). Plus précisément, les halos sont significativement parallèles entre eux, et la direction de plan qu'ils définissent statistiquement n'est pas distinguable de celle de (μ) .

Par ailleurs, il arrive souvent qu'un même halo soit commun à plusieurs galaxies (spirales ou irrégulières), qui constituent un SYSTEME aplati - et que la direction de ce système manifeste le même parallélisme avec (μ).

Il est significatif que l'hypothèse du parallélisme soit d'autant mieux vérifiée que les observations sont plus précises; en particulier pour les trois spirales du GROUPE LOCAL: la nôtre d'abord, pour laquelle l'angle de gauchissement est particulièrement grand (80°), puis M31 (Andromède) et M33 . Notre Galaxie et M31 possèdent chacune un système de galaxies naines satellites, systèmes qui sont dans les deux cas TRES NETTEMENT PARALLELES à (μ) .

Comment interpréter ces faits? Il est possible que le halo d'une galaxie spirale ne soit qu'une partie d'un nuage quasi-plan de grande dimensions (une "FEUILLE"), constituant la configuration primitive où la galaxie et ses annexes éventuelles ont pu apparaître par condensation.

Dans notre environnement ces feuilles semblent groupées par paquets de tailles diverses, constituant le Groupe Local et le Super-Amas Local; ces objets présentent donc une structure "feuilletée".

La direction de ce feuilletage est PARALLELE à la zone (μ) ; s'il ne s'agit pas d'une coïncidence accidentelle concernant notre environnement seulement, on peut envisager une STRATIFICATION HIERARCHISEE de la matière dans tout l'Univers.

Il est possible au contraire que l'effet observé soit un alignement des halos galactiques SUR LE SUPERA-AMAS ENVIRONNANT, que ces amas soient orientés au hasard, et que ce ne soit que par coïncidence que le nôtre soit parallèle à (μ).

C'est difficile à trancher pour l'instant, parce que les amas répertoriés sont mal connus; les hasards de la perspective y jouent un rôle non négligeable, comme dans les constellations traditionnelles.

UN MODELE D'UNIVERS CONFRONTE AUX OBSERVATIONS

Nous savons que, dans diverses directions, la répartition des distances présente de grands trous: ceci indique l'existence de structures spatiales bien délimitées; mais les mesures de distance sont encore trop rares pour que nous puissions décrire ces structures sans géocentrisme.

------§15------

Une éventuelle stratification pourrait aussi s'observer à grande distance, par les RAIES D'ABSORPTION FINES dans les spectres des quasars; et particulièrement par les "FORETS LYMAN α ", que l'on observe pour la plupart des quasars lointains ($z > 2$). Il s'agit de raies d'absorption fines, intenses et nombreuses (typiquement 30 à 50 par objet), qu'on interprète comme résultat de l'absorption Lyman α par des nuages interposés composés d'hydrogène peu enrichi en éléments lourds. La grande fréquence de ces interpositions est évidemment plus facile à interpréter s'il s'agit de FEUILLES du type (14) que de halos sphéroïdaux.

------§16------

On a pu mesurer la CINEMATIQUE COLLECTIVE des galaxies autour de nous (vitesse et apex du groupe local par rapport à son environnement, par exemple). Il se trouve que les vitesses vectorielles obtenues sont, à la précision des mesures, PARALLELES A LA STRATIFICATION GENERALE. Interprétation de ce fait: la cinématique de la matière, là où nous la connaissons, est compatible avec la PERMANENCE de la stratification.

A l'échelle des feuilles (§14) , l'existence d'une telle agitation tangentielle donnerait une explication dynamique de la ROTATION et du GAUCHISSEMENT des galaxies (par conservation des trois composantes du moment cinétique lors du processus de condensation).

------§17------

L'observation précise du RAYONNEMENT COSMOLOGIQUE permet de mesurer la vitesse de la Terre par rapport au "référentiel du rayonnement"; par composition des vitesses, on peut donc atteindre la "CINEMATIQUE ABSOLUE" des galaxies.

Là encore on constate que LES VITESSES SONT SITUEES DANS UN MEME PLAN, dont la direction COINCIDE AVEC CELLE DE (μ) avec la précision des mesures.

A N N E X E S

ANNEXE DU PARAGRAPHE 1.

UN MODELE D'UNIVERS CONFRONTE AUX OBSERVATIONS

La relativité générale est fondée sur les hypothèses suivantes:

a) Le champ de gravitation est caractérisé par le tenseur métrique d'espace-temps $g_{\mu\nu}$: une masse d'épreuve GRAVITE selon une géodésique de cette métrique. (Les 10 composantes de ce tenseur s'appellent les POTENTIELS DE GRAVITATION; dans les calculs, nous choisirons la signature (+ − − −)).

b) Le TENSEUR IMPULSION-ENERGIE $T_{\mu\nu}$, est astreint à avoir une divergence riemannienne nulle - ce qui s'exprime par 4 équations dites "de conservation". Il constitue la SOURCE de la gravitation; ou, si l'on préfère, ses composantes $T_{\mu\nu}$ sont les dix sources du champ de gravitation.

c) Les équations de la gravitation vérifient le "principe de relativité générale"; en langage géométrique, elles doivent être invariantes par l'action du groupe des difféomorphismes de la variété espace-temps.

A partir de ces hypothèses, convenablement précisées, on établit que la forme la plus générale possible pour les équations de champ est la suivante:

$$T_{\mu\nu} = -A\, g_{\mu\nu} + B\, (R_{\mu\nu} - 1/2\, R\, g_{\mu\nu}) + \text{(invariants du second degré)} + \ldots \tag{1*1}$$

$R_{\mu\nu}$ désignant le tenseur de Ricci, R la courbure contractée. Quant aux coefficients A, B,..., ce sont des constantes universelles sur lesquelles on ne peut rien dire a priori, et qui doivent donc être MESUREES.

Un peu d'ANALYSE DIMENSIONNELLE va nous être utile. Bien que ce soit un usage courant, il est tout-à-fait dépourvu de sens (sauf conventions cachées et révocables) de faire figurer la constante c dans les équations de la relativité générale; on ne peut choisir que deux unités fondamentales (masse et longueur par exemple), et la troisième (ici le temps) est dérivée (exemples: année = année lumière; s = 2.99792459 E10 cm). L'analyse dimensionnelle correcte de la relativité générale s'obtient en attribuant la dimension L^2 au tenseur covariant $g_{\mu\nu}$, donc L^{-2} au tenseur contravariant $g^{\mu\nu}$ (descendre ou monter les indices augmente ainsi l'exposant de L de ± 2), 1 au tenseur de Ricci $R_{\mu\nu}$, donc L^2 à la courbure contractée R. En remarquant que la masse spécifique et la pression sont des composantes du tenseur mixte T^{μ}_{ν}, on trouve l'équation aux dimensions

$$\left[T_{\mu\nu} \right] = M\, L^{-1} \tag{1*2}$$

d'où, par homogénéité de l'équation (1*1), les équations aux dimensions:

(1*3) $$\left[A \right] = M\,L^{-3}, \quad \left[B \right] = M\,L^{-1}.$$

Ceci montre que la constante A sera prépondérante à grande échelle, B à échelle "moyenne"; quant aux termes suivants ($M\,L$, $M\,L^{3}$,...), qui n'ont d'ailleurs pas été détectés, ils ne pourraient être perceptibles qu'à échelle microscopique, et par conséquent il est raisonnable de les prendre nuls en cosmologie.

Diverses méthodes d'interprétation des constantes A et B sont possibles, en particulier la construction des solutions exactes à symétrie sphérique généralisant celle de Schwarzschild. Nous nous contenterons de donner ici l'APPROXIMATION NEWTONIENNE de l'équation (1*1), à savoir les équations de Poisson modifiées

(1*4) $$\overrightarrow{\text{rot}}\ \vec{g} = 0 \qquad \text{div}\ \vec{g} = -\,4\pi G \quad + \Lambda$$

où l'on a posé

(1*5) $$G = 1/(8\pi B), \qquad \Lambda = A / B$$

avec donc les équations aux dimensions

(1*6) $$\left[G \right] = L\,M^{-1}, \quad \left[\Lambda \right] = L^{-2}$$

On reconnait donc dans G la constante de Newton

(1*7) $$G = 7.4243\ E\text{-}29\ \text{cm g}^{-1}$$

ce qui fixe la valeur de B . Quant à l'influence du terme Λ, supposé positif pour fixer les idées, on constate sur l'équation (1*4) qu'elle consiste en un EFFET REPULSIF DU VIDE; qu'un milieu homogène pourra rester en équilibre gravitationnel si sa masse spécifique est égale à $\Lambda/4\pi G = 2A$. Tous ces résultats se retrouvent QUANTITATIVEMENT en effectuant le traitement relativiste; en particulier l'équilibre d'un fluide de densité 2A correspond au premier modèle statique construit par Einstein.

Il est clair que la constante Λ (ou A) ne sera appréciable que par des observations à TRES GRANDE ECHELLE; il est donc indispensable pour pouvoir la mesurer de se placer à l'échelle cosmologique, en comparant à l'observation une solution des équations d'Einstein (1), écrites aussi:

(1*8) $$R_{\mu\nu} - 1/2\ R\ g_{\mu\nu} - \Lambda\ g_{\mu\nu} = 8\pi G\ \ T_{\mu\nu}$$

UN MODELE D'UNIVERS CONFRONTE AUX OBSERVATIONS

Pour chercher cette solution nous pouvons utiliser aujourd'hui l'observation du rayonnement cosmologique; plus précisément, le fait que ce rayonnement ne diffère pas sensiblement du rayonnement de Planck du corps noir. On ajoute un ingrédient, à savoir le "principe cosmologique" selon lequel ce fait n'est pas seulement vrai ICI et AUJOURD'HUI; un raisonnement physico-géométrique simple (voir Souriau 1974) permet de montrer qu'il IMPLIQUE le suivant:

(1*9) Le quadrivecteur-température de Planck $\vec{\beta}$ du rayonnement (direction: "référentiel" commun de la matière et du rayonnement; longueur: 1/kT; sens: futur) est Killing-conforme pour la métrique (la dérivée de Lie $L_{\vec{\beta}}\, g_{\mu\nu}$ est proportionnelle à $g_{\mu\nu}$),

dont une interprétation physique est la suivante:

(1*10) Les échanges d'énergie, d'impulsion, d'entropie entre le rayonnement et la matière sont négligeables.

Ces résultats ne mettent en jeu que les propriétés passives de la matière et de la lumière dans un espace riemannien (principe des géodésiques).

Il est clair que cette hypothèse (1*9) ne doit être prise que comme approximation de travail, valable seulement à grande échelle, extra-galactique par exemple. Sa conséquence (1*10) en montre les limites: nous savons bien, par exemple, que le rayonnement à 3°K agit sur les molécules interstellaires à l'intérieur de la Galaxie, donc que (1*10) y est en défaut. Ces réserves faites, nous allons constater qu'elle est SUFFISANTE pour déterminer la classe des modèles cosmologiques.

Introduisons MAINTENANT les équations de champ (1*8), prenant donc en compte les SOURCES du champ de gravitation. Parmi celles-ci, les CONTRAINTES, qui sont des composantes T^{μ}_{ν}, sont évidemment les mieux connues: nous avons de bonnes raisons de penser que cette contrainte est, quasiment partout et toujours, une PRESSION pure (autrement dit, que le cisaillement est une source négligeable du champ de gravitation!). Compte tenu de l'hypothèse (1*9), on en DEDUIT (Loc. Cit.) l'ISOTROPIE et l'HOMOGENEITE de l'espace; de façon précise, on obtient une métrique de Robertson:

(1*11) $$ds^2 = dt^2 - R(t)^2\, d\sigma^2$$

où t est un paramètre (le temps universel bien entendu) et où $d\sigma$ est la métrique d'une variété riemannienne de dimension 3 à COURBURE CONSTANTE (l'"espace", sur lequel les galaxies sont fixes); les variétés de ce type ont été classées par Riemann lui-même (du point de vue local; la classification globale est donnée dans Wolf (1967)); leur propriété fondamentale est de posséder un GROUPE D'ISOMETRIES de dimension 6 , dont l'existence exprime l'ISOTROPIE et l'HOMOGENEITE de l'Univers.

Les propriétés du spectre du rayonnement à 3°K IMPLIQUENT donc l'isotropie de la répartition de matière dans le ciel - et la constatation de ce fait par les décomptes de galaxies apparaît comme une vérification de la cohérence du modèle.

la fonction R(t) (à ne pas confondre avec la courbure contractée) est inconnue a priori; elle est proportionnelle à la température réciproque β = 1/(kT) ; nous la normaliserons en la prenant égale à 1 à l'époque actuelle, soit

(1*12) $$R = T_{\circ} / T \quad (T_{\circ} \# \ 2.7°K).$$

L'expansion (croissance de R) s'accompagne donc d'un refroidissement. On montre aussi que le REDSHIFT z d'un objet observé est donné par

(1*13) $$R = 1/(1+z)$$

R étant la valeur de R(t) à la date où l'objet est observé.

On considère aussi la "fonction de Hubble"

(1*14) $$H(t) = 1/R \ dR/dt$$

qui mesure le taux d'expansion (dimensionnellement, $[H] = L^{-1}$), et le paramètre de décélération q , sans dimensions, défini par

(1*15) $$1+q = d(1/H)/ dt$$

Pour pouvoir intégrer les équations, il faut évaluer la pression; dans l'Univers déjà "homogénéisé" par les approximations consenties, il semble admissible de considérer que le seul terme non négligeable soit la PRESSION DE RADIATION dûe au rayonnement cosmologique, à savoir $[\pi^2/45M^3]\ (kT)^4$. Alors les équations s'intègrent - et du même coup nous connaissons LES DIX COMPOSANTES $T_{\mu\nu}$ qui figurent au second membre. Ce tenseur est obtenu par superposition de celui du RAYONNEMENT (tel qu'il est défini par la théorie de Planck) et de celui d'une POUSSIERE (fluide parfait de pression négligeable). Nous connaissons donc l'évolution de la masse spécifique de cette poussière sans avoir fait sur elle aucune hypothèse préalable.

Cette circonstance est tout-à-fait étrangère à la physique newtonienne: les équations de la gravitation ont pu être intégrées sans connaître la répartition des masses. L'utilisation de la relativité introduit donc une différence conceptuelle RADICALE, dûe évidemment au fait qu'il y a dix équations de la gravitation au lieu d'une.

Détaillons le formulaire des résultats. La fonction R(t) est définie par une intégrale elliptique

(1*16) $$t = \frac{1}{H_\circ} \int \frac{R\,dR}{\sqrt{P(R)}}$$

où P désigne un polynôme du 4ème degré

(1*17) $$P(R) = \alpha_\circ + \Omega_\circ R - k_\circ R^2 + \lambda_\circ R^4 ;$$

$H_\circ$ est le paramètre de Hubble, valeur actuelle de $H(t)$ (1*14); il est habituel de le repérer par le paramètre sans dimension h tel que

(1*18) $$H = h \times 100 \text{ Km/s/Mpc} = h \times 1.081 \text{ E-28 cm}^{-1} = h/(2998 \text{ Mpc})$$

rappelons que les valeurs proposées contradictoirement par Sandage et De Vaucouleurs sont respectivement $h = 0.55$ et $h = 0.95$.

Il résulte de (1*16) que

(1*19) $$H = H_\circ \sqrt{P(R)} / R^2$$

et

(1*20) $$q = 1 - R\,P'(R)/ 2P(R)$$

- On sait que $R = 1$ à l'époque actuelle (1*12); la formule (1*16) montre que la valeur actuelle de P est 1, donc que

(1*21) $$\alpha_\circ + \Omega_\circ - k_\circ + \lambda_\circ = 1;$$

ces coefficients du polynôme P sont des nombres sans dimension, que nous allons interpréter.

Le terme radiatif $\alpha_\circ$ est connu par la loi de Planck du corps noir, et sa valeur est

(1*22) $$\alpha_0 = 8\pi^3 G\,(kT_0)^4 / (45\,\hbar^3\,H_0^2) \quad \# \quad 5E\text{-}5$$

valeur qui pourrait être augmentée d'un facteur voisin de 1.5 si on voulait tenir compte des éventuels neutrinos thermiques - à condition qu'on les suppose de masse nulle (Weinberg 1978). De toutes façons ce terme est petit dans la partie observée de l'Univers; il devient au contraire prépondérant dans l'extrapolation vers le big-bang (en négligeant les autres termes, on obtient le modèle radiatif pur de Gamow). Le modèle de FRIEDMANN - LEMAITRE proprement dit consiste dans l'approximation = 0; pour le traitement des quasars, il donne donc des résultats très proches de ceux du modèle complet.

- Le PARAMETRE DE DENSITE Ω_0 est relié à la masse spécifique actuelle ρ_0 par

(1*23) $$\Omega_0 = 8\pi G \rho_0 / 3H_0^2$$

la masse spécifique de la matière à tout instant étant donnée par

(1*24) $$\rho = \rho_0 / R^3$$

- La COURBURE REDUITE k_0 indique par son signe la géométrie de l'espace (Lobatchevski, Euclide ou Riemann selon que k_0 est <0, nul ou >0); attention! k_0 N'EST PAS NORMALISE aux valeurs 0 ou 1 ; sa valeur numérique sera utilisée plus loin (1*33,1*38).

- Enfin la CONSTANTE COSMOLOGIQUE REDUITE λ_0 est donné par

(1*25) $$\lambda_0 = \Lambda / (3\,H_0^2)$$

Puisque ces 4 paramètres sont reliés par la relation (1*21) et que l'un d'eux est connu (1*22), ils s'expriment au moyen de deux d'entre eux; l'usage est de choisir le paramètre de densité Ω_0 et la valeur actuelle q_0 du paramètre de décélération q ; soit, grâce à (1*20)

(1*26) $$q_0 = 1 - \Omega_0 / 2 + k_0 - 2\lambda_0$$

d'où

(1*27) $$\lambda_0 = \Omega_0 / 2 - q_0 + \alpha_0$$

(1*28) $$k_0 = (3/2)\Omega_0 - q_0 + 2\alpha_0 - 1$$

Enfin (1*16) donne:

(1*29) $$\text{AGE DE L'UNIVERS} = \frac{1}{H_0}\int_{1/(1+z)}^{1} \frac{R\,dR}{\sqrt{P(R)}}$$

- Il est remarquable que la COSMOLOGIE NEWTONIENNE, que l'on obtient en joignant les équations d'Euler d'un fluide à pression négligeable aux équations de Poisson modifiées ci-dessus (1*4) conduise EXACTEMENT aux mêmes résultats (avec $\alpha_0 = 0$, comme dans le modèle de Friedmann-Lemaître). Evidemment le coefficient k_0 ne s'interprète plus comme une "courbure spatiale", mais comme une intégrale première dont la valeur dépend des conditions initiales (c'est l'intégrale des forces vives appliquée aux trajectoires des molécules).

Revenons à la relativité, et étudions la répartition SPATIALE des objets.

La DISTANCE GEODESIQUE d'un objet de redshift z (mesurée à l'époque actuelle, ce qui évidemment ne correspond à aucune procédure expérimentale) est égale à

(1*30) $$r = \frac{1}{H_0}\int_{1/(1+z)}^{1} \frac{dR}{\sqrt{P(R)}}$$

Pour localiser cet objet dans l'espace on peut, comme dans les formules de Robertson - Walker, utiliser cette distance r et des paramètres angulaires. Mais, dans le cas qui va nous intéresser où la courbure k_0 est positive, il sera beaucoup plus commode de procéder autrement.

En effet, la classification de Riemann montre que l'espace est alors une sphère S3 (localement; quelques hypothèses globales permettent des développements mathématiques pittoresques qui pour l'instant restent sans vérification possible; voir Wolf 1967). Nous placerons donc les points matériels (la Terre et les différents quasars) sur la sphère mathématique décrite dans l'espace $\mathbb{R}^4$ des matrices-colonnes

(1*31) $$X = \begin{pmatrix} x_1 \\ x_2 \\ x_3 \\ x_4 \end{pmatrix}$$

par l'équation

(1*32) $$x_1^2 + x_2^2 + x_3^2 + x_4^2 = 1$$

et nous aurons

(1*33) $$d\sigma^2 = (dx_1^2 + dx_2^2 + dx_3^2 + dx_4^2) / (k_\circ H_\circ^2)$$

Bien entendu la symétrie de la sphère nous permettra d'y situer arbitrairement la terre T ; nous choisirons

(1*34) $$T = \begin{pmatrix} 1 \\ 0 \\ 0 \\ 0 \end{pmatrix}$$

Notons qu'une DIRECTION D DANS LE CIEL se traite comme un vecteur unitaire tangent à S3 au point T ; nous pourrons poser

(1*35) $$D = \begin{pmatrix} 0 \\ \cos\delta \ \cos\alpha \\ \cos\delta \ \sin\alpha \\ \sin\delta \end{pmatrix}$$

α et δ étant respectivement l'ascension droite et la déclinaison (1950) de l'objet. Nous connaissons alors ses coordonnées quadridimensionnelles par la formule

(1*36) $$X = T \cos\tau + D \sin\tau$$

ou encore

(1*37) $$X = \begin{pmatrix} \cos\tau \\ \sin\tau \ \cos\delta \ \cos\alpha \\ \sin\tau \ \cos\delta \ \sin\alpha \\ \sin\tau \ \sin\delta \end{pmatrix}$$

UN MODELE D'UNIVERS CONFRONTE AUX OBSERVATIONS

τ étant la distance ANGULAIRE sur S3 de l'objet à la Terre; soit (voir (1*30,*33))

(1*38) $$\tau = r\, H_\circ \sqrt{k_\circ} = \sqrt{k_\circ} \int_{1/(1+z)}^{1} \frac{dR}{\sqrt{P(R)}}$$

Si on possède un catalogue de quasars (comportant les positions célestes et les redshifts), et si on a choisi les paramètres cosmologiques, ces formules permettent de calculer les coordonnées quadri-dimensionnelles de chaque objet; on possède ainsi, sous forme de fichier, la répartition des quasars dans l'espace sphérique S3 . Cette répartition peut s'utiliser pour tous les tests statistiques que l'on voudra.

(1*39) Considérons une sphère Sn , de rayon unité, plongée dans l'espace numérique Rn+1 ; si on la projette orthogonalement sur un (n-1)-plan diamétral, on obtient évidemment une (n-1)-boule (boule = intérieur de sphère) Bn-1; si V est une région intérieure à cette boule, l'ensemble des points de Sn dont la projection appartient à V a un n-volume égal au produit par 2π du (n-1)-volume de V. Ce résultat a été établi il y a 2200 ans par Archimède dans le cas n = 3 (l'aire d'une calotte sphérique est égale à 2π fois sa hauteur), ce qui lui a permis de trouver l'aire 4π de la sphère S2 ; dans le cas n = 3 , on voit que le volume de S3 vaut $2\pi^2$. Ceci permet d'évaluer la MASSE TOTALE de l'Univers, compte tenu de sa densité actuelle :

(1*40) $$M = 2\pi^2 \rho_\circ / (H_\circ \sqrt{k_\circ})^3 ;$$

en utilisant (1*23), il vient

(1*41) $$M = \frac{3\pi\Omega_\circ}{4\, k_\circ^{3/2}} \quad \frac{1}{G\, H_\circ}$$

soit M # = 1.2E80 atomes d'hydrogène.

Mais le théorème d'Archimède a une autre utilité: il va nous servir à VISUALISER la répartition spatiale des objets. Il nous montre en effet que des points équipartis aléatoirement dans l'espace S3 se projettent selon des points équipartis du disque B2 ; si tous les quasars étaient connus, leurs projections sur B2 devraient donc présenter une densité constante - quel que soit le 2-plan de projection choisi.

Il est intéressant de choisir un plan de projection passant par la Terre T (1*34); on construit cette carte en choisissant un autre point K de S3 qui soit orthogonal à T, et en représentant chaque quasar par le point de coordonnés (T,X) , (K,X) , X ayant la valeur (1*37) [la notation (,) désigne le produit scalaire à

4 dimensions]. C'est ainsi qu'on a obtenu la figure 1, carte globale de l'Univers.

Des objets situés à une même distance de T (dans l'espace S3) se projettent sur une même corde orthogonale au diamètre TT' ; on lit donc directement sur la figure l'effet de sélection par la distance (ou par le redshift, ou encore par la date à laquelle l'objet est observé). Ce qu'on observe le plus nettement sur la figure, c'est d'une part l'accumulation des objets visibles au voisinage de la Terre T ; d'autre part le cut-off bien connu aux distances correspondant à des redshifts plus grands que 3.53 . On répète souvent (par exemple Osmer 1982) que les quasars étaient un millier de fois plus fréquents dans une période primitive qu'ils ne le sont aujourd'hui; l'examen de la figure montre que cette interprétation des données n'est pas obligatoire: elle repose sur le choix d'un autre modèle d'Univers.

Mais on voit autre chose sur cette figure 1: la ZONE D'ABSENCE annoncée; elle se distingue sur le diamètre horizontal - malgré les difficultés dues à l'échelle. On l'appréciera mieux sur les figures 2 a,b,c.

La zone est encore visible, bien que tordue, sur la figure (2a), qui est construite en adoptant le modèle naïf de Hubble (espace euclidien, distance proportionnelle au redshift); elle a meilleure mine sur (2b), qui utilise le modèle relativiste et des coordonnées géodésiques. Nous avons distingué en noir sur (2c) les objets figurant dans le catalogue de Burbidge et al. (1977) à partir duquel nous avons publié les premiers résultats de cette étude (Souriau 1980). Depuis cette date, le nombre d'objets qui ont été observés dans cette région a augmenté de 150 % ; et pourtant LES NOUVEAUX OBJETS (ronds blancs) ONT RESPECTE LA ZONE INITIALE. Un tel fait serait hautement improbable s'il s'agissait d'une coïncidence.

On peut aussi se demander s'il ne s'agit pas d'un effet de sélection par l'observation, si par exemple certains redshifts ne sont pas plus difficiles à observer que d'autres. Mais comme le manque d'objets se manifeste sur une bande étroite de redshifts ($\Delta z/z$ # 1/20) DONT LE CENTRE DEPEND FORTEMENT DE LA DIRECTION DE VISEE (de z = 1 à z = 2.8), cette explication est bien difficile à soutenir.

Un élément essentiel pour l'étude de significativité est une bonne connaissance de la précision sur la MESURE DES REDSHIFTS. On constate qu'un spectre individuel, même à basse résolution (Δz = 1/100), conduit à des conclusions significatives, mais que les résultats provisoires obtenus par la méthode du prisme objectif ou du grism sont insuffisants: l'imprécision sur le redshift est du même ordre que la largeur de la zone à observer (quand il ne s'agit pas d'une simple confusion avec une étoile!).

Nous ne développerons pas davantage ici la discussion: le lecteur trouvera une étude détaillée des diverses éventualités de coïncidences, d'effets de sélection ou d'artefacts concernant cette zone d'absence dans les 250 pages de la thèse de Roland Triay (1981).

Le point de départ de notre travail est un CATALOGUE DE Q.S.O. (Triay, mars 1982) adapté au problème, c'est-à-dire aussi complet que possible d'une part, et d'autre part critique en ce qui concerne la fiabilité et la précision des redshifts. Le catalogue publié comporte 1840 objets, dont 1206 possédant des spectres individuels

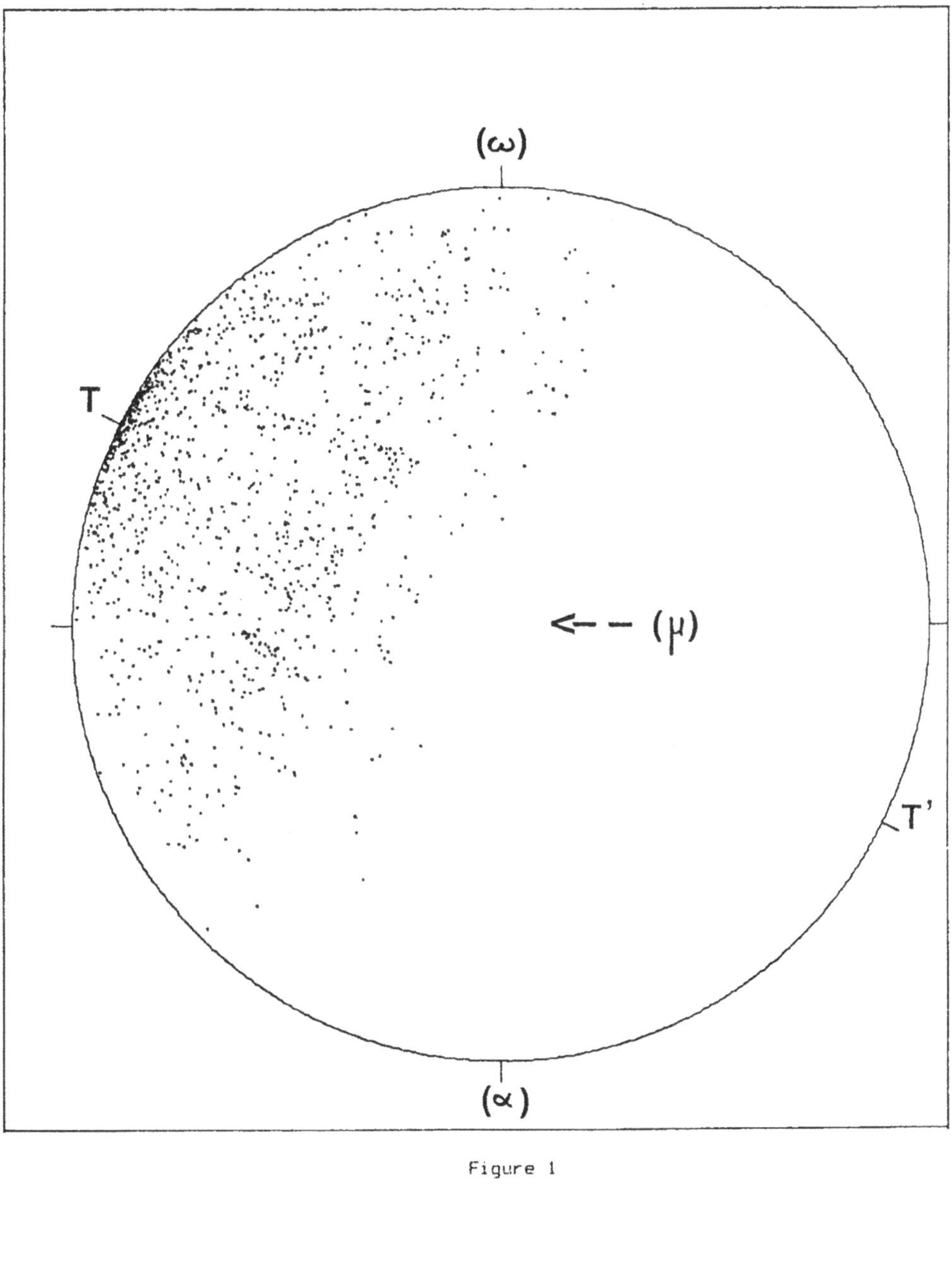

Figure 1

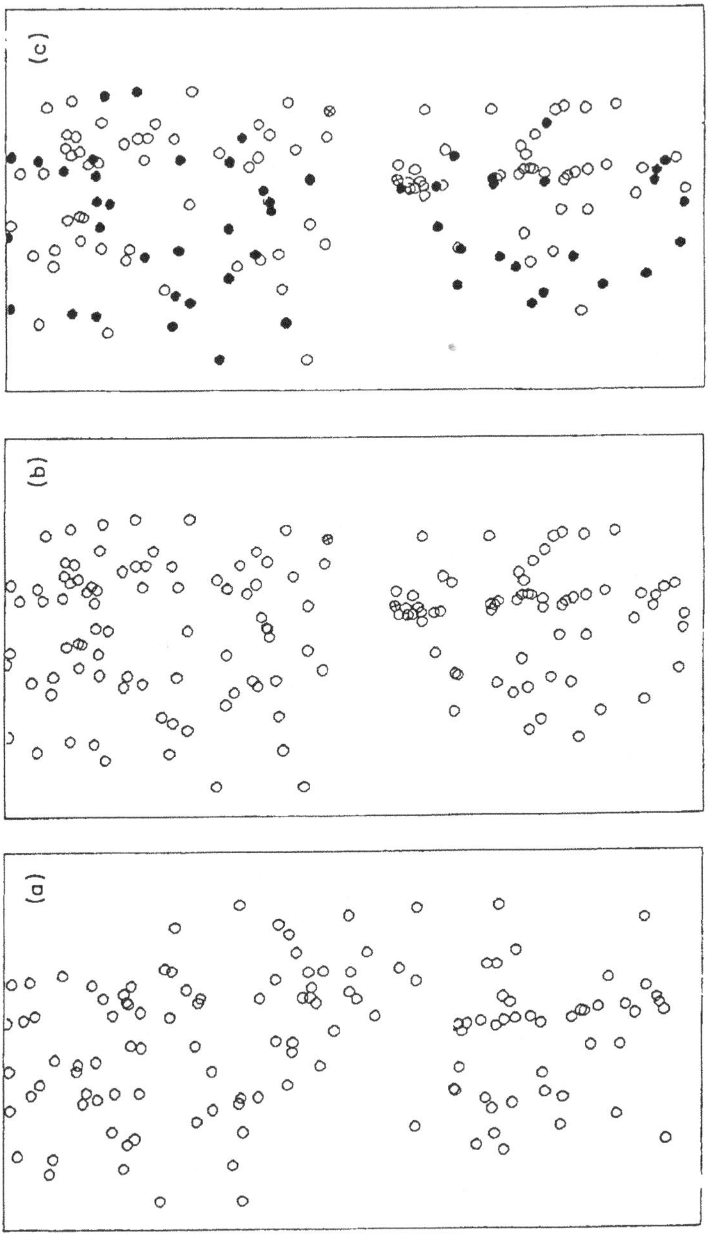

Figure 2

confirmés. Constamment mis à jour, il comporte aujourd'hui 1900 objets (août 1982).

Grâce à ce catalogue, nous avons pu indiquer (Fliche et al. 1980) une liste d'objets que leurs déterminations provisoires au prisme objectif plaçaient DANS LA ZONE (μ) ou à proximité immédiate; lorsque des spectres individuels ont été publiés, ils se sont obligeamment répartis de part et d'autre (ils font partie des objets marqués sur la figure 2).

Une deuxième liste de 6 objets douteux figure dans Fliche et al. (1982 I); les 3 premiers ont déjà été réobservés par Peterson et Savage (Peterson 1982); il en résulte que deux sont situés sur les bords de (μ) (ils sont indiqués par le signe $\otimes$ sur la figure 2), et que le troisième n'est qu'une étoile.

Une troisième liste plus étendue (9 objets dans (μ); 14 dans chaque bande adjacente de même largeur) est soumise aux observateurs (Fliche et al. 1982 III).

Il est clair que l'existence de cette zone d'absence impose des contraintes très strictes aux paramètres cosmologiques. Un balayage systématique sur ceux-ci (avec des programmes d'optimisation écrits en Fortran V et en Pascal) donne une estimation des valeurs compatibles avec ces contraintes:

(1*42) Ω_0: compris entre 0.05 et 0.20

(des valeurs plus élevées sont possibles, mais moins satisfaisantes);

(1*43) $q_0 = -1.12 \pm 0.01$;

- coordonnées équatoriales (1950) de l'axe perpendiculaire à (μ), dans la direction opposée:

(1*44) (ω): (5h 45mn 30s $\pm$ 1mn, +6° 50' $\pm$ 15');

ce "pôle cosmique" est situé dans le quadrilatère d'Orion, à 2° environ de Betelgeuse;

- plus courte distance de la Terre au plan médian de (μ) :

(1*45) $d_0 = (0.905 \pm 0.01)\ c/H_0 = (2700 \pm 30)$ Mpc/h

- épaisseur de la zone (μ):

(1*46) $\lesssim$ 120 Mpc/h

(1*47) Pour éviter tout choix subjectif, nous avons déterminé un modèle de travail en maximisant, sur les données disponibles, le rapport volume(μ) / volume(Univers), qui atteint ainsi la valeur 1/75 ; la valeur correspondante des paramètres est: Ω_0 = 0.1015 ; q_0 = -1.1226 ; k_0 = 0.2750 ; λ_0 = 1.1735 ; d_0 = 0.9073 ; latitude cosmique de la Terre: 0.4758 = 27°15' ; coordonnées équatoriales du pôle (ω) en radians: 1.5072 , +0.1198 . Ces chiffres, qui ne sont évidemment pas tous significatifs, permettent de vérifier les divers calculs. Voici par exemple une table, calculée avec ce modèle, qui donne la zone de redshifts occupée par (μ) en fonction de la distance angulaire (en degrés) au centre (α) (1*44):

(1*48)

0°	:	de	z= 0.849	à	0.887
5°	:	de	z= 0.852	à	0.890
10°	:	de	z= 0.860	à	0.898
15°	:	de	z= 0.874	à	0.913
20°	:	de	z= 0.894	à	0.934
25°	:	de	z= 0.921	à	0.962
30°	:	de	z= 0.956	à	0.998
35°	:	de	z= 1.000	à	1.044
40°	:	de	z= 1.054	à	1.100
45°	:	de	z= 1.121	à	1.170
50°	:	de	z= 1.203	à	1.256
55°	:	de	z= 1.305	à	1.363
60°	:	de	z= 1.433	à	1.496
65°	:	de	z= 1.593	à	1.665
70°	:	de	z= 1.799	à	1.881
75°	:	de	z= 2.065	à	2.162
80°	:	de	z= 2.415	à	2.534
85°	:	de	z= 2.881	à	3.033
90°	:	de	z= 3.504	à	3.702
95°	:	de	z= 4.321	à	4.583

Cette table permet d'évaluer facilement la position d'un quasar par rapport à (μ); ainsi 1213-003, situé dans Virgo à 83.0±0.3° de (α) est observé au bord le plus proche de (μ) avec le redshift z = 2.684.

ANNEXE DU PARAGRAPHE 2.

La chronologie du modèle est donnée par les formules (1*16), (1*13) et en particulier (1*29); nous allons donner aux coefficients du polynôme P (1*17) les valeurs du modèle standard (1*47).

Si deux événements survenant au même point sont observés aux redshifts z1 et z2, ils sont séparés par un intervalle de temps

(2*1) $$\Delta t = \frac{1}{H_\circ} \int_{1/(1+z2)}^{1/(1+z1)} \frac{R\, dR}{\sqrt{P(R)}}$$

L'intégrale est facile à calculer par la méthode de Simpson; on trouve ainsi, pour l'âge de l'Univers, la valeur:

(2*2) $A = 1.64 / H_\circ = 16.1 / h$ milliards d'années;

Dans le cas du modèle à CONSTANTE COSMOLOGIQUE NULLE ayant la même valeur du paramètre de densité $\Omega_\circ$ (ce qui entraîne $k_\circ = -0.90$), on trouve par la même formule

(2*3) $A' = 0.90 / H_\circ = 8.8 / h$ milliards d'années.

Par ailleurs les estimations proposées pour les âges maximum des étoiles et des amas globulaires sont de l'ordre de 14 à 16 milliards d'années (voir Tammann et al. 1979). Si on choisit $\Lambda = 0$, on constate donc que h est liée par la contrainte

(2*4) $h < 0.55$

et que par conséquent la valeur de Sandage 0.55 est JUSTE CELLE QU'IL FAUT pour pouvoir adopter le modèle traditionnel $\Lambda = 0$.

Par contre, avec le présent modèle, la valeur $h = 0.95$ de De Vaucouleurs est compatible avec ces contraintes, puisqu'elle conduit à l'âge

(2*5) $A = 17$ milliards d'années

qui implique que la formation de la Galaxie et de ses premières structures stellaires est relativement primordiale; corrélativement, la valeur $h = 0.55$ pose quelques questions de temps perdu.

Il ne faut pas oublier les erreurs probables sur toutes les données manipulées ici; mais les variations possibles des paramètres du modèle (voir (1*42,43)) ne modifient pas sensiblement ces conclusions.

ANNEXE DU PARAGRAPHE 3.

Comme nous l'avons vu, les valeurs du PARAMETRE DE DENSITE $\Omega_\circ$ compatibles avec l'existence de (μ) s'étendent de 0.05 à 0.20 et même un peu au delà; la valeur du modèle standard (1*47) étant très proche de 0.1 .

Il se trouve que ces valeurs sont à peu près celles que l'on rencontre dans la littérature; citons par exemple Gunn et Tinsley (1975): 0.06 ± 0.03, Peebles (1979): 0.4 ± 0.2; Gunn (1978): 0.08 .

On peut évaluer Ω_0 par la dynamique des amas de galaxies; cette méthode a l'avantage de prendre en compte toutes les formes de matière même invisible (par exemple les halos galactiques, les galaxies naines non détectables directement, les éventuels trous noirs, neutrinos massifs et autres monopoles qui pourraient circuler dans l'amas) et de fournir Ω_0 indépendamment de la valeur de H_0 ; deux difficultés: les observations sont rares et leur interprétation délicate.

Notons un problème particulier: le THEOREME DU VIRIEL que l'on utilise dans ces évaluations repose sur l'existence d'un groupe d'invariance pour la loi de Newton:

(3*1) $$t \longrightarrow s^3 t, \qquad \vec{r} \longrightarrow s^2 \vec{r} \qquad (s > 0)$$

(cf. la troisième loi de Képler); or ce groupe disparait quand la loi de gravitation est modifiée par la prise en compte de la constante cosmologique (voir ci-dessus (1*4)).

Le modèle standard, grâce aux formules (1*25, 1*47), donne la valeur de Λ:

(3*2) $$\Lambda = (1700 \text{ Mpc})^{-2} = (5.5 \text{ E } 9 \text{ années})^{-2}$$

ce qui est évidemment négligeable à l'échelle des amas; par contre la densité d'équilibre associée (Cf.(1*4))

(3*3) $$\Lambda / (4\pi G) = 4 \text{ E } -29 \text{ g cm}^{-3}$$

est tout-à-fait typique d'une densité d'amas. Le rôle éventuel de la constante cosmologique dans l'évaluation dynamique de Ω_0 reste donc à préciser.

Il est évidemment possible qu'il existe de la matière non détectée située entre les amas - et même entre les super-amas, qui pourrait affecter très fortement la valeur de Ω_0 ; mais les évaluations proposées restent très arbitraires.

Le modèle étudié ne semble donc pas en désaccord avec l'état de nos connaissances sur la densité moyenne de l'Univers.

UN MODELE D'UNIVERS CONFRONTE AUX OBSERVATIONS

ANNEXE DU PARAGRAPHE 4.

La relation REDSHIFT - LUMINOSITE pour les galaxies peut se visualiser par le diagramme de Hubble - qui pour les petites valeurs de z ne permet que de vérifier la loi de Hubble.

Pour obtenir par cette voie une estimation du paramètre q_o , il faut:

a) obtenir un échantillon homogène de galaxies dont les redshifts s'étendent assez loin;

b) avoir une idée précise de la correction K (par suite du redshift, les observations s'étendent vers la partie ultraviolette du spectre d'émission; il faut savoir comparer la luminosité bolométrique d'objets qui sont observés dans des fenêtres spectrales différentes); ou obtenir des mesures extra-atmosphériques;

c) Corriger les mesures du fait que la partie du disque prise en compte diminue quand la distance augmente;

d) pouvoir évaluer l'effet d'un effet d'évolution statistique des objets (les observations s'étendent vers le passé).

Ces corrections b), c), d) correspondent à des effets systématiques du même ordre que l'effet à mesurer; b) et c) sont en principe accessibles, mais l'évaluation directe de l'effet d'évolution d) reste assez arbitraire: dans un premier stade, il est raisonnable de se contenter d'examiner si une évolution négligeable est compatible avec les faits. Sur un échantillon prélevé dans les amas d'Abell proches, Hoessel et al.(1980) trouvent ainsi

$$q_o = -0.55 \pm 0.45$$

compatible donc avec la valeur proposée ici.

ANNEXE DU PARAGRAPHE 5.

La même méthode s'applique en remplaçant les galaxies par les QUASARS. Les grandes valeurs de leurs redshifts excluent évidemment l'emploi de méthodes utilisant des approximations valables au voisinage de z = 0 .

Comme dans le cas des galaxies, se pose le problème de la correction K et celui de l'évolution; mais il n'y a évidemment pas d'effet de disque.

En attendant des mesures extra-atmosphériques en nombre suffisant, la correction K est possible si le spectre ultra-violet des objets est suffisamment homogène et s'il est connu. Nous avons utilisé à cet effet une méthode colorimétrique dont le principe est du à Sandage - qui nous a fourni un SPECTRE COMPOSITE DES QUASARS (fig. 3) qui est en accord remarquable avec le spectre de 3C 273, obtenu directement par tir de fusée. A noter l'apparition des principales raies du spectre, qui sortent directement de l'analyse colorimétrique statistique. L'EFFET DE RAIE dans

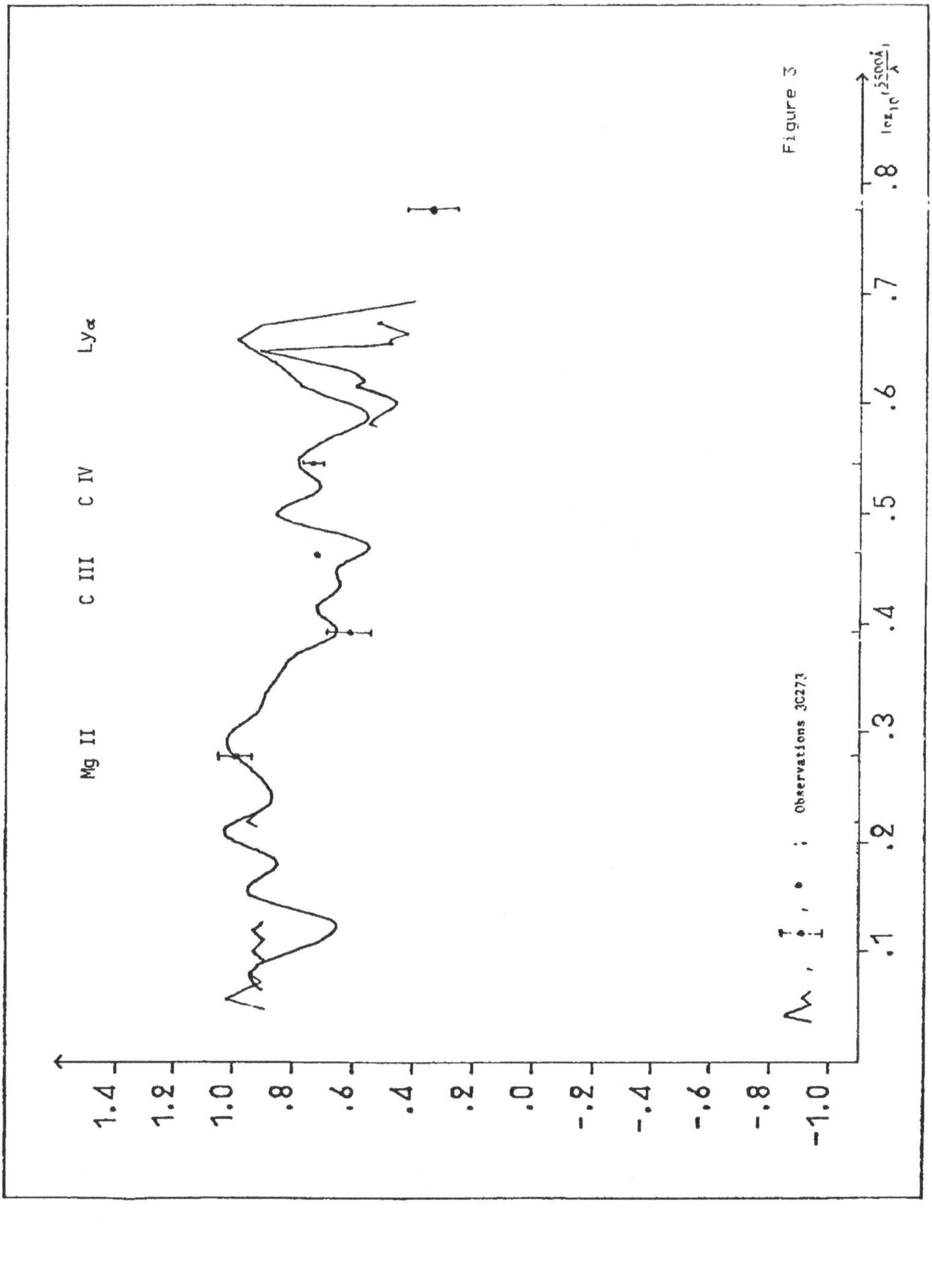
Mg II
C III
C IV
Lyα
Observations 3C273
1.4
1.2
1.0
.8
.6
.4
.2
.0
-.2
-.4
-.6
-.8
-1.0
.1
.2
.3
.4
.5
.6
.7
.8

Figure 3

la correction K est assez important, et par conséquent les formules empiriques supposant que ce spectre est de type synchrotron sont complètement erronées.

La netteté du spectre composite (obtenu avec des données colorimétriques s'étendant jusqu'à z = 3.53), sa similitude avec le spectre de 3C273, quasar proche, indiquent une grande unité de la physique du phénomène quasar, et permettent donc d'envisager l'hypothèse de non-évolution statistique de ces objets.

Pour une étude détaillée, nous renvoyons à Fliche et al.(1979) et à Fliche (1981, thèse de Doctorat), qui conduisent aux modèles

(5*1) $$\Omega_0 = 0.08 \pm 0.04, \quad q_0 = -1.10 \pm 0.04$$

en accord avec le modèle standard ci-dessus.

Cet accord est visualisé sur la figure 4 . Les objets pour lesquels on dispose de données photométriques sont représentés par un cercle, dont la surface est proportionnelle à la puissance émise (calculée dans le modèle et avec la correction K ci-dessus); il ne s'agit donc pas d'une échelle logarithmique. On constate que les objets les plus brillants sont extrêmement homogènes.

Les seuls effets évolutifs qu'on puisse noter sont:

a) le confinement des objets dans un demi-cercle, correspondant à l'apparition des objets 2.5 / h milliards d'années après le big-bang (voir l'interprétation au §5); cette date est calculée avec la formule (2*1), le modèle (1*47) et z = 3.53 (redshift record de OQ172).

b) un manque de quelques objets brillants proches; mais on ne peut affirmer que cet effet soit significatif.

On voit que l'évolution en luminosité des objets n'est pas plus significative que leur évolution en nombre; la possibilité d'annuler SIMULTANEMENT ces deux effets par la prise en compte du seul paramètre Λ est un fait important pour l'interprétation.

ANNEXE DU PARAGRAPHE 6.

Puisque ce test est simplement satisfaisant sans être très significatif, nous renvoyons aux travaux suivants: Fliche et al (1979), Fliche (1981).

ANNEXE DU PARAGRAPHE 7.

Le chemin parcouru sur S3 par la lumière, entre le moment du découplage (z = 1100) et l'époque actuelle est donné par la formule (1*38); avec le modèle (1*47), le calcul numérique de l'intégrale donne

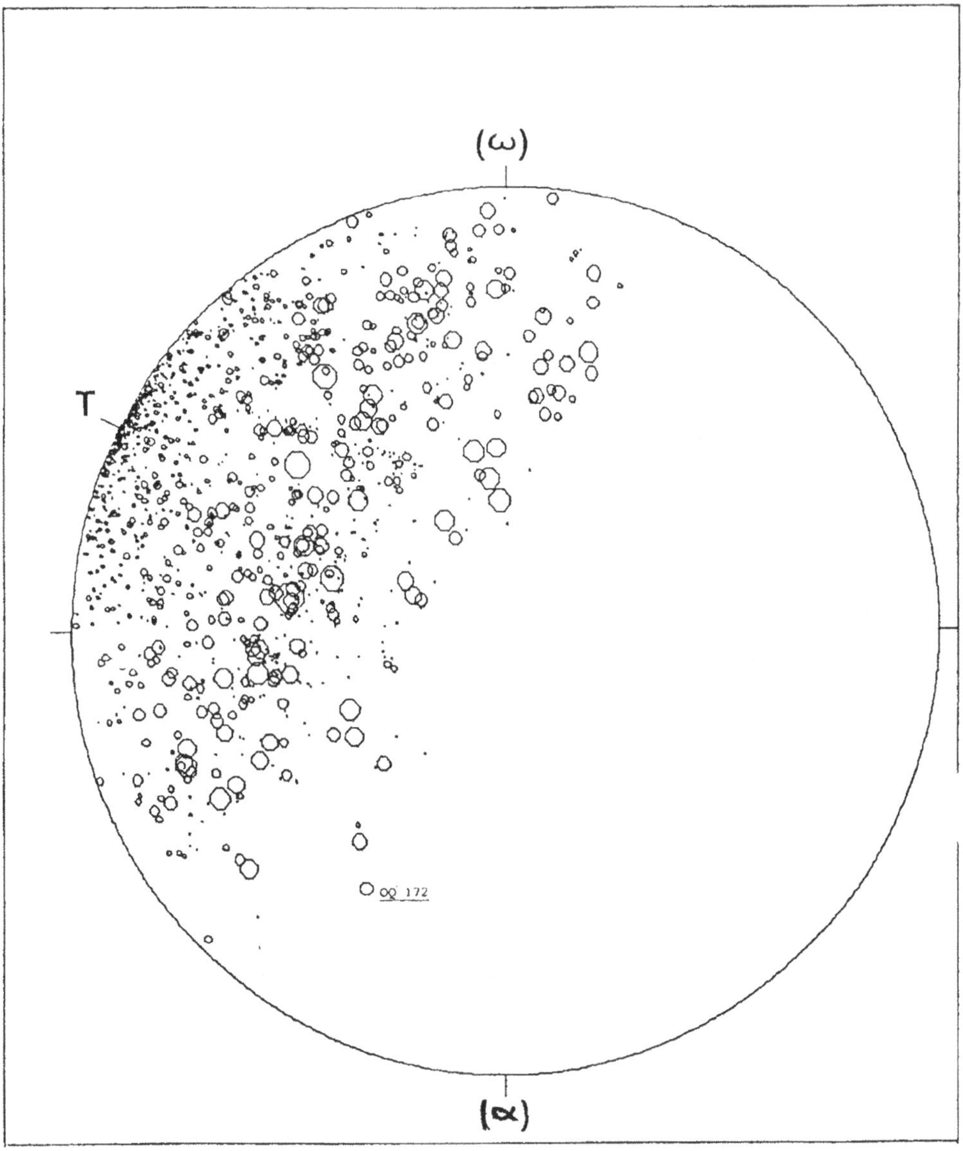

Figure 4

UN MODELE D'UNIVERS CONFRONTE AUX OBSERVATIONS

(7*1) $$\tau = 3.165 = 181.3°$$

soit très légèrement plus d'un demi-tour. Il en résulte que les sources observées du rayonnement à 3 °K sont originaires d'une très petite région (1/400 000 ème de l'Univers). Ces chiffres n'ont qu'une valeur d'exemple parmi les modèles compatibles avec l'existence de (μ) .

ANNEXE DES PARAGRAPHES 9 ET 10.

Pour l'étude de la physique de la région de contact matière - antimatière nous renvoyons à Schatzman (1982).

ANNEXE DU PARAGRAPHE 11.

(11*1) Nous avons vu en (1*11) que la symétrie d'un modèle de Friedmann est caractérisée par un GROUPE G1 (ici, le groupe des rotations de la sphère S3); mais les phénomènes que nous étudions (existence de (μ) , répartition de matière et d'antimatière), ont une symétrie moindre, définie mathématiquement par le SOUS-GROUPE G2 des rotations de S3 qui conservent (μ)); ou, ce qui revient au même, qui laissent fixes les deux pôles (α) et (ω).

Nous sommes dans une situation appelée SYMETRIE BRISEE, dont il existe de nombreux exemples en physique et en astrophysique. De même une planète, en équilibre sous sa propre gravitation, possède la SYMETRIE SPHERIQUE; sous l'effet du mouvement diurne, seule subsiste la SYMETRIE DE REVOLUTION. Le groupe G1 est ici consitué des rotations autour du centre, le sous-groupe G2 des rotations autour de l'axe des pôles.

(11*2) Dans cette situation, la notion mathématique d'ORBITE va jouer un rôle important: on appelle ainsi les ensembles de points que l'on peut échanger par l'action d'un groupe.

Dans l'exemple planétaire, les orbites (O1) de G1 sont les sphères concentriques; pour G2 , les orbites (O2) sont les cercles parallèles axés sur la ligne des pôles. Par construction, CHAQUE (O1) EST UNE REUNION DE (O2).

Même dans le cas où on ignore le mécanisme intime d'un phénomène, la symétrie nous permet des prédictions sur ses résultats - qui s'expriment en termes d'orbites. Ainsi les transitions de phase géophysiques sont, au niveau 1, des sphères concentriques; au niveau 2, elles sont perturbées en réunions de (O2), donc en surfaces de révolution. La météorologie pousse la brisure de symétrie au stade qualitatif, et partage l'atmosphère en zones composées de (O2); telles les bandes de Jupiter.

La classification en orbites est donc essentielle pour l'analyse de tous les processus physiques; elle impose d'ailleurs le choix des paramètres pertinents pour cette étude.

Dans le modèle cosmologique étudié, les orbites (O1) sont indexées par le TEMPS COSMOLOGIQUE t et, pour chaque date, constituent l'ESPACE S3 tout entier). Cet espace tridimensionnel est "feuilleté" par les orbites (O2), qui sont les

SURFACES PARALLELES à (μ) - y compris (μ) elle-même. Chaque (O2) peut se repérer par une "latitude cosmique" (qui vaut 27° environ pour la Terre, nous l'avons dit).

Que pouvons-nous en déduire sur le résultat des mécanismes cosmogoniques? Au niveau 1, un espace homogène, dont les caractéristiques évoluent en fonction du temps - le modèle de Friedmann-Lemaître lui-même. Au niveau 2, le stade qualitatif permet l'existence de transitions sur des orbites (O2); la transition la plus évidente est la discontinuité matière -- antimatière à la traversée de (μ); mais des transitions de l'état de la matière (de son degré de condensation par exemple) sont possibles sur d'autres (O2); ce qui conduit à envisager une STRATIFICATION DE L'ESPACE, dans la direction parallèle à (μ).

(11*3) Au stade quantitatif, on doit aussi envisager de perturber le modèle lui-même. La théorie des groupes joue encore un rôle fondamental dans ce travail; ainsi l'interprétation des résultats de Lifchitz et Khalatnikov (1963) doit-elle être nuancée: certaines variables prises en compte ne sont pas invariantes par le groupe de jauge gravitationnelle, et n'ont donc pas de signification physique (voir la thèse de Fliche (1981)).

(11*4) Enfin notre position même d'observateur se manifeste par une réduction des symétries: G1 se réduit au groupe G'1 des rotations autour de la Terre: le niveau 1 implique un ciel isotrope. Le niveau 2 prévoit une ANISOTROPIE du ciel, définie par le groupe G'2 des rotations du ciel autour des deux points (α) et (ω); les orbites de G'2 sont des cercles parallèles, indexées par une "déclinaison cosmique" Bc qui vaut -90° au pôle (ω), +90° en (α); l'intersection du plan Bc = 0 avec le plan galactique sera pris comme origine pour la "longitude cosmique" Lc ; ce qui conduit aux formules

$$\cos Bc \cos Lc = \cos b \cos(\ell - 109.4)$$
$$\cos Bc \sin Lc = \cos b \sin(\ell - 109.4) \cos(79.3) + \sin b \sin(79.3)$$
$$\sin Bc = \sin b \cos(79.3) - \cos b \sin(\ell - 109.4) \sin(79.3)$$

reliant les coordonnées galactiques ℓ , b aux coordonnées cosmiques Lc , Bc; ces coordonnées permettent de construire des cartes du ciel adaptées à la recherche de l'anisotropie éventuelle (figures 5 et 12); pour des raisons qui vont apparaître, elles sont proches des coordonnées supergalactiques B , L définies par De Vaucouleurs (remplacer 109.4 par 137.3 et 79.3 par 83.7).

(11*5) La STRATIFICATION que nous avons envisagée peut se manifester tout d'abord dans la répartition spatiale des QUASARS. On distingue sur la figure 2 , à côté de l'équateur (μ) , quelques bandes horizontales, qui peuvent indiquer des ZONES D'ABSENCE DE QUASARS, parallèles à (μ). Une étude statistique de ces zones se trouve dans Souriau (1980) et Triay(1981). Comme elles sont beaucoup moins larges que (μ), elles sont plus sensibles à l'imprécision des mesures de redshift; avec les données disponibles, leur existence semble probable (voir l'annexe 12).

ANNEXE DU PARAGRAPHE 12.

(12*1) Les spectres des quasars lointains présentent en général des RAIES D'ABSORPTION nombreuses; elles sont pour la plupart très fines. Mais il y a quelques exemples de raies au contraire TRES LARGES (comme dans les spectres des super-novae; mais l'analogie en reste là; en particulier elles ne manifestent pas d'évolution sensible depuis qu'on les observe).

Deux types d'interprétation de ces raies larges sont proposés: phénomène INTRINSEQUE (produit par le quasar émissif) ou INTERPOSITION de matière quelque part sur la ligne de visée.

Dans l'interprétation intrinsèque, le blue-shift par rapport à l'émission et la largeur de ces raies se traduisent en termes de VITESSES - qui sont considérables (quelques dixièmes de c). En cas d'interposition, la largeur indique plus vraisemblablement l'EPAISSEUR du nuage observé (qui serait à l'échelle cosmologique) qu'une agitation relativiste.

Le prototype du quasar à raies larges est PHL5200, pour lequel l'hypothèse intrinsèque est la plus fréquemment admise; cependant on n'est pas parvenu à la concrétiser par un modèle plausible (voir Junkkarinen et al. 1982). Dans le cas de Q1246-057, très analogue, le décalage de redshift émission-absorption conduit à éliminer l'hypothèse intrinsèque.

(12*2) Considérons d'abord quatre de ces objets, qui présentent une raie Lyman α large en absorption: A: (1246-057), B: (1331+170), C: (1334+285) ou RS23, D: (2225-055) ou PHL5200. Si on considère qu'ils indiquent des nuages interposés, on peut les situer dans l'espace par les mêmes formules que les quasars eux-mêmes.

Avec le modèle euclidien de Hubble, on constate que ces objets sont presque coplanaires; plus précisément, que la sphère qu'ils déterminent a un très grand rayon (# 5 c/H_0) et que nous sommes situés à l'intérieur. Le point le plus proche est situé à z = 0.9, dans la direction

(17h 45mn, -6° 40')

qui coïncide avec (α) (voir (1*44)). Cette sphère n'est autre que (μ) , et par conséquent A , B , C , D POURRAIENT S'INTERPRETER COMME DES NUAGES SITUES DANS LA ZONE (μ) DEPOURVUE DE QUASARS. Dans le scénario de Schatzman (§9), on peut envisager que le retard au découplage ait été suffisant dans (μ) pour empêcher la formation de galaxies, mais pas celle de nuages de grandes dimensions.

Si on utilise le modèle de travail (1*47), l'appartenance de ces nuages à (μ) peut se vérifier par le calcul de leur latitude cosmique (voir la formule (1*37) et l'annexe 11); on trouve

Objet	z (abs.)	Lat. cosmique
A	2.05	+9'
B	1.78	-57'
C	1.87	-31'
D	1.88	-50'

Ces valeurs posent un problème, parce que (μ) s'étend entre -36' et +36' seulement. Mais avant de jeter l'interprétation, on peut prendre en compte plusieurs possibilités: variations permises du modèle; imprécision sur les redshifts; on peut envisager aussi les perturbations de 2ème niveau du modèle évoquées en (11*3), ainsi que des perturbations de 3ème niveau qui pourraient déformer (μ) (nous comparons à 20 Mpc prés la position par rapport au modèle d'objets dont les distances mutuelles dépassent 10 000 Mpc). Mais il se peut que l'obtention de spectres de très grande qualité ou la découverte de nouveaux objets (par observation systématique des "antiquasars") renouvelle la question.

(12*3) On connaît une dizaine d'autres objets à raies larges, qui NE SONT PAS SITUES DANS L'EQUATEUR (μ). Mais on constate, sur 6 d'entre eux, que les nuages correspondants sont situés DANS L'UNE DES AUTRES ZONES D'ABSENCE PRESUMEES (11*5); leur liste est donnée dans Fliche et al.(1982 I). Deux d'entre eux, 2240-370 et 2238-412 , écartés de 4° , présentent le même redshift d'absorption 1.70 et pourraient peut-être indiquer une même structure, observée sur plus de 200 Mpc.

ANNEXE DU PARAGRAPHE 13.

L'anisotropie éventuelle du ciel a été expliquée à l'annexe 11; les coordonnées cosmiques qui y sont définies sont utilisées pour dresser la carte du ciel (fig. 5), où figurent les "outstanding galaxies" définies par De Vaucouleurs (1975): galaxies les plus proches, les plus brillantes ou les plus grandes dans le ciel, limitées à la distance de 10 Mpc. On constate que la région $-30° < Bc < 30°$, qui constitue la moitié du ciel, contient 53 galaxies, contre 6 à l'extérieur; ce qui est hautement significatif. L'usage des coordonnées supergalactiques conduit à un rapport moins favorable, 48 contre 11 (de Vaucouleurs 1975).

Le "Nuage Local" constitué de ces objets est donc une structure très aplatie, et pratiquement parallèle à (μ). Il contient le "Groupe Local", composé d'une dizaine d'objets proches ($d < 1$ Mpc), et qui présente donc la même disposition.

Le "Super-Amas Local", s'étendant jusqu'à 30 Mpc, a été aussi défini par De Vaucouleurs (1953). Il présente la même structure aplatie, et la direction adoptée pour son axe (pôle supergalactique) diffère de 28° de (α). Récemment Tully (1982) y a détecté un feuilletage par des structures qui pourraient être analogues au Groupe Local. Il serait intéressant de reprendre la statistique de ces observations en libérant la direction du pôle, pour savoir si l'écart avec (α) est significatif.

ANNEXE DU PARAGRAPHE 14.

Etudions l'ORIENTATION DU HALO des trois spirales du Groupe Local.

Les 12 satellites connus de NOTRE GALAXIE (en particulier les Nuages de Magellan LMC et SMC) sont portés sur la figure 6 , établie en COORDONNEES COSMIQUES (voir (11*4)). Ils sont TOUS situés dans la bande $-30° < Bc < 30°$, et par conséquent le Système Galactique est lui aussi aplati et parallèle à (μ).

Sur la même carte sont portés d'autres éléments probables du halo, les NUAGES H I à GRANDE VITESSE (d'après De Vaucouleurs et al.1975). Même disposition, notamment pour le "flot Magellanique" qui part de LMC et SMC et qui s'étend sur un tiers de l'équateur cosmique.

Le PLAN GALACTIQUE, selon sa définition classique, est porté sur cette carte. On

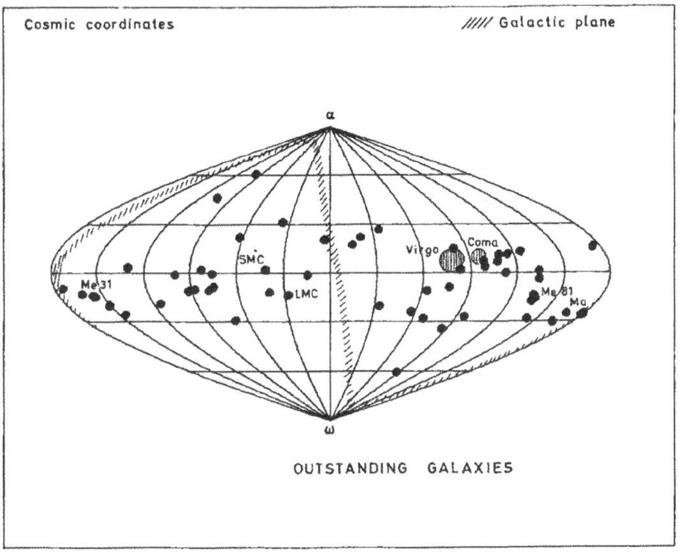

Figure 5

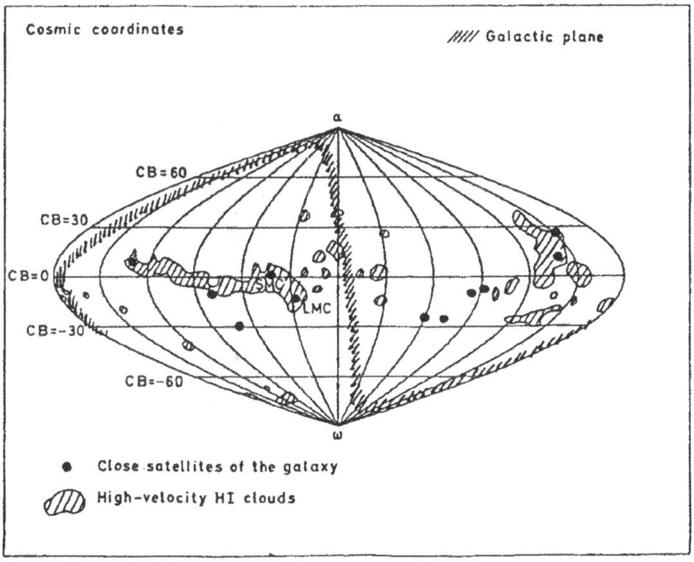

Figure 6

constate que la disposition du halo n'est pas un effet de sélection par l'obscurcissement galactique; l'angle entre ce plan galactique et le plan du halo (80°) donne une mesure du GAUCHISSEMENT de notre Galaxie.

En ce qui concerne la Nébuleuse d'Andromède" M31 , le gauchissement est visible sur la figure 7 , composition d'une image optique et d'un contour radio, due à Emerson et al. (1978). Nous avons simplement marqué la direction parallèle à (μ) - visiblement en accord avec la direction de gauchissement telle qu'elle résulte du contour radio.

Les satellites probables de M31 sont portés sur la figure 8 , ainsi que la direction de (μ). Là aussi on voit une structure plate et parallèle à la direction marquée, qui s'étend sur près de 30° du ciel. Une partie en est probablement cachée par le plan Galactique (δ > 50°).

Le gauchissement de M33 a été étudié par Sandage et Humphrey (1980), à partir des données optiques. Sur la figure 9 , l'évolution du "sommet du grand axe" a été indiquée par ces auteurs (ligne interrompue); nous avons indiqué la droite parallèle à (μ), l'accord est excellent.

Sandage et Humphrey estiment d'autre part à 0.42 le rapport d'aplatissement b/a pour les régions externes; en supposant qu'il s'agit de cercles parallèles à (μ) vus en perspective, le calcul donne b/a = 0.44.

La figure 10 donne les cartes radio de M33 d'après Huchtmeier (1978): courbes de densité et de vitesse. La direction de (μ) semble aussi valable que le "Major Axis" proposé par cet auteur.

En ce qui concerne les autres galaxies, nous renvoyons à Fliche et al. (1982 II), où se trouvent des cartes analogues concernant notamment la "nébuleuse des Chiens de chasse" M51, M83, M101, M81, NGC 4490/85.

Ce travail contient aussi une étude des échantillons de halos étudiés par Rots (1980) et Bosma (1981). A 90 %, ils sont situés dans la moitié du ciel -30° < Bc <30°.

Pour chaque objet qui apparaît allongé sur le ciel, la direction de cet allongement est repérée par l'angle de position (p.a.); Δp.a. désigne la différence entre la valeur de p.a. DONNEE PAR CES AUTEURS et celle qui se calcule par l'hypothèse de parallélisme à (μ). Les histogrammes sont les suivants:

Echantillon de Rots

Δ p.a.:	0°-15	15-30	30-45	45-60	60-75	75-90°
sur 33:	11	2	9	3	4	4

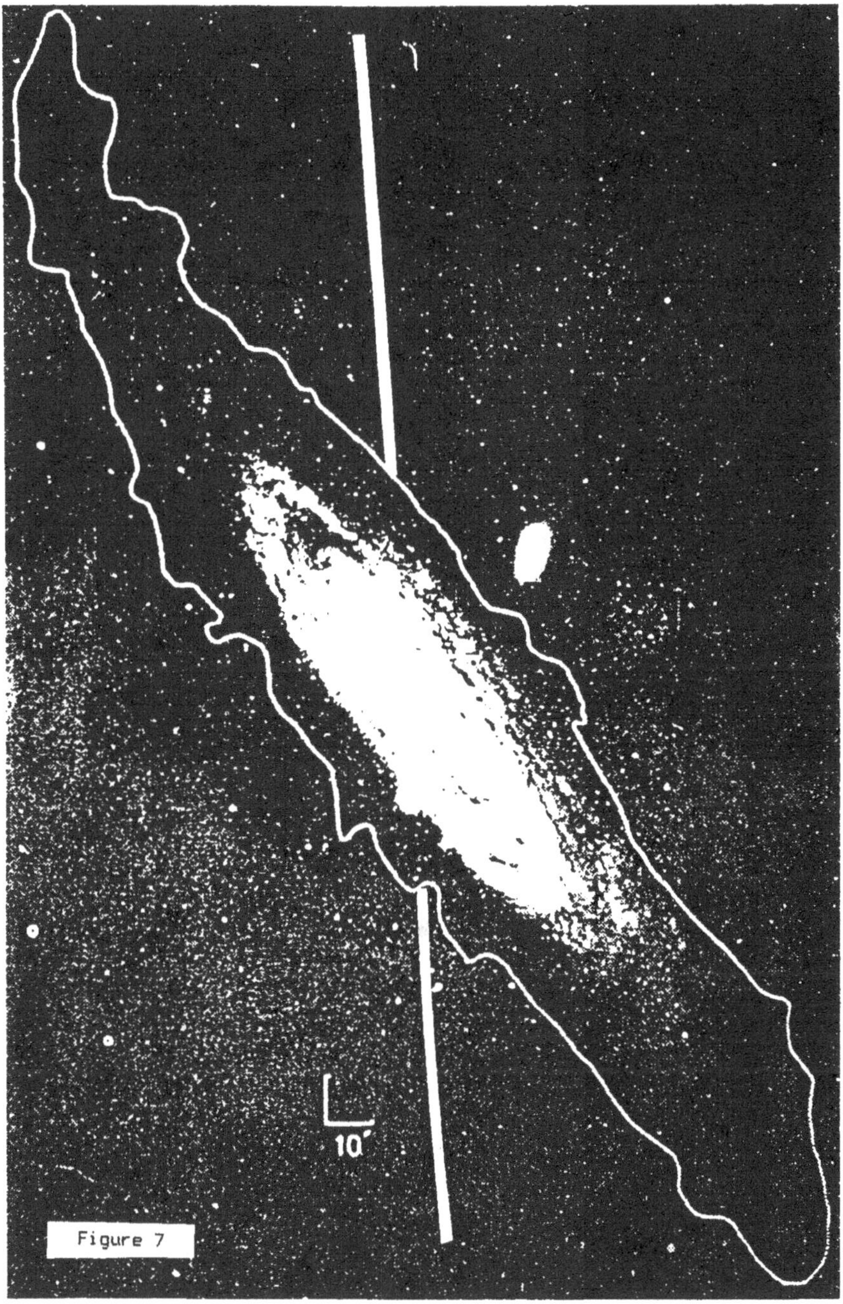

Figure 7

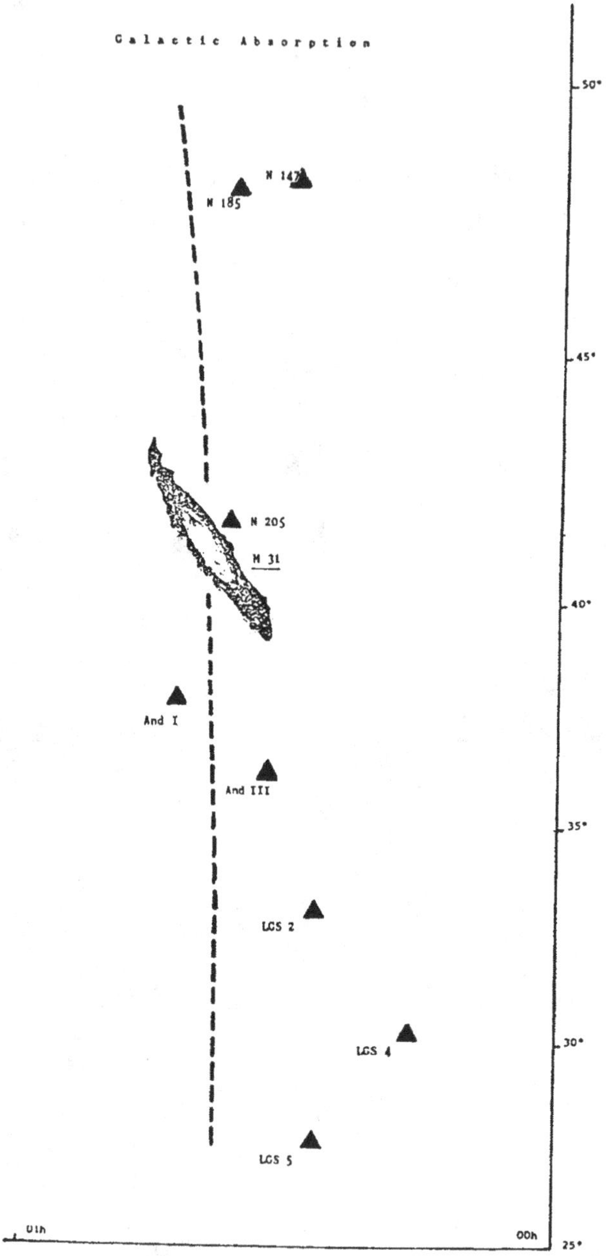
Galactic Absorption
50°
M 147
M 185
45°
M 205
M 31
40°
And I
And III
35°
LGS 2
LGS 4
30°
LGS 5
01h
00h
25°

Figure 8

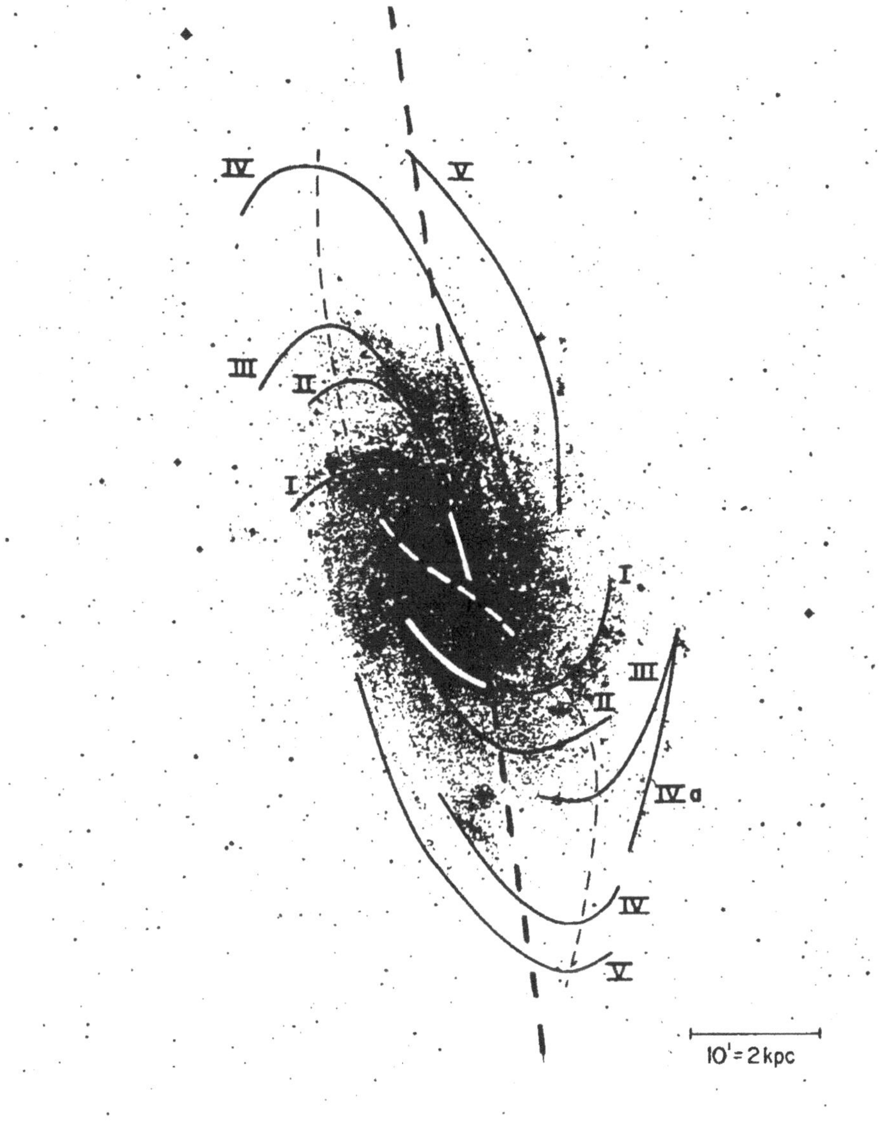

Figure 9

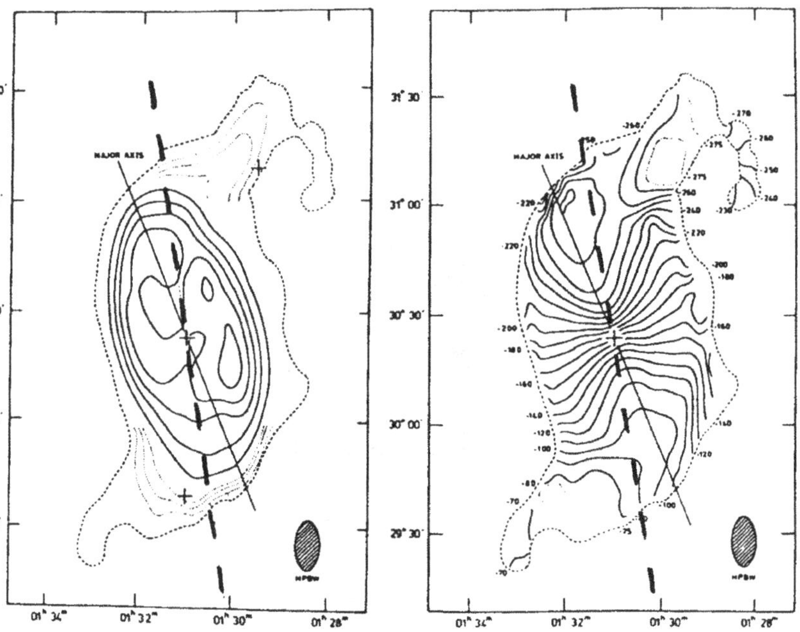
MAJOR AXIS
31° 30'
31° 00'
30° 30'
30° 00'
29° 30'
01h 34m
01h 32m
01h 30m
01h 28m
HPBW

Figure 10

UN MODELE D'UNIVERS CONFRONTE AUX OBSERVATIONS

Echantillon de Bosma

p.a.:	0°-15	15-30	30-45	45-60	60-75	75-90°
sur 28:	11	4	5	2	4	2

Le lecteur pourra vérifier que ces histogrammes sont très significatifs.

Une autre méthode pour apprécier cette corrélation consiste à prolonger par un grand cercle du ciel le petit axe apparent de chaque halo observé; et à pointer les intersections deux à deux de ces cercles. Pour éviter un effet de sélection par la position des objets, on supprime les points qui sont trop près des deux galaxies (à moins de 30°).

La corrélation étudiée doit avoir pour effet de resserrer ces points autour de (α) et de (ω). C'est ce qu'on constate sur la carte 11A (échantillon de Bosma); le témoin 11B est obtenu en remplaçant chaque angle de position par un angle tiré au sort. (N.B.: ces cartes ne sont pas disposées comme les précédentes: (α) est au centre, (ω) apparaît aux deux extrémités).

L'examen de la région centrale de 11A montre que l'effet ne peut pas être amélioré sensiblement en choisissant une autre direction pour les pôles.

Rappelons que l'hypothèse du parallélisme à (μ), même rigoureuse, N'ENTRAINE PAS l'alignement des angles de position, mais seulement une forte corrélation (voir le §14); c'est exactement ce qu'on observe.

ANNEXE DU PARAGRAPHE 15.

Sur les raies d'absorption fines des Q.S.O., et particulièrement sur la forêt Lyman α ,voir Sargent et al. (1980), Sargent et al. (1982).

ANNEXE DES PARAGRAPHES 16 ET 17.

Nous allons interpréter la cinématique des référentiels suivants:

(S): le Soleil;

(L): une moyenne prise sur le Groupe Local. La valeur adoptée pour la vitesse relative (S/L), donnée par De Vaucouleurs & Peters (1981,I), est une moyenne de 5 estimations dues à divers auteurs.

(G): un échantillon de 300 galaxies, étudiées par De vaucouleurs et al.(1981 II), dont les distances sont comprises entre 2 et 29 Mpc, et qui sont réparties dans tout le ciel. Les distances sont évaluées par la largeur de la raie à 21cm (méthode adaptée de Tully et Fisher) .

(G'): Un échantillon de 200 galaxies, observées optiquement, traitées par De Vaucouleurs et Peters (1981 I).

(G"): Un échantillon de 84 galaxies observées dans les deux hémisphères à Jodrell Bank et à Parkes; la vitesse relative (SG") est donnée par Hart & Davies (1982).

11A

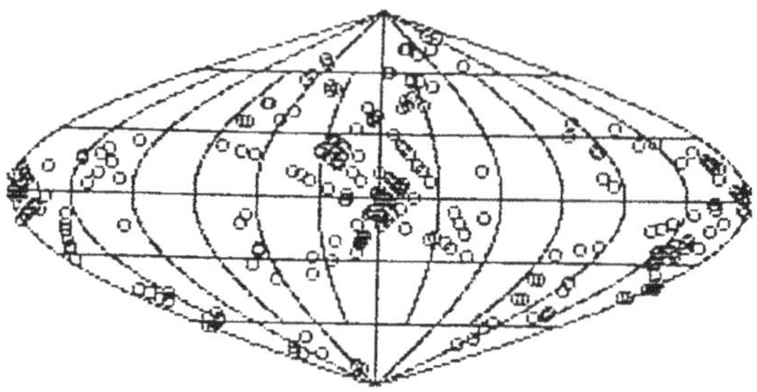

11B

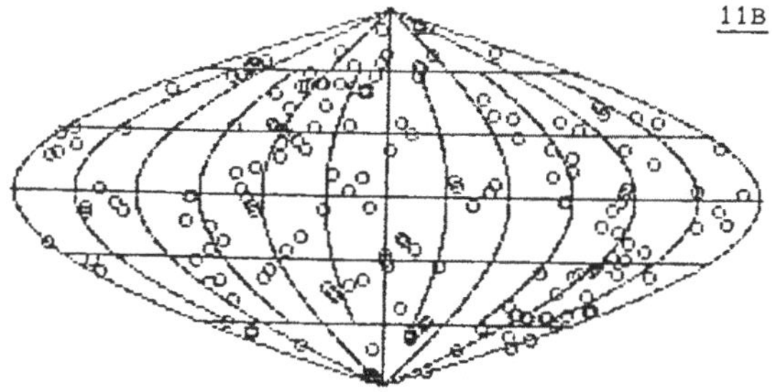

Figure 11

(*): Le référentiel du rayonnement cosmologique à 3°K. La vitesse relative (S/*) est une moyenne des observations suivantes: Smoot et al. (1977); Corey (1978); Gorenstein (1978); Cheng et al. (1979); Cheng et al. (1980), pondérée par De Vaucouleurs et al.(1981 II).

La figure 12, en coordonnées cosmiques, comporte les 6 apex relatifs des 4 référentiels (S), (L), (G), (*). On constate que ces points SONT TOUS SITUES PRATIQUEMENT SUR L'EQUATEUR, donc qu'ils déterminent une cinématique plane parallèle à la zone (μ) .

L'ensemble des résultats cinématiques se lit sur la table suivante, qui donne la vitesse "absolue" (c'est-à-dire relative au rayonnement (*)) des divers référentiels ci-dessus - telles qu'elles sont fournies par leurs auteurs. Les vitesses sont exprimées en km/s, dans des coordonnées rectangulaires cosmiques:

	X	Y	Z	
(S)	-182	317	-5	(±20)
(L)	-504	414	-2	(±30)
(G)	-464	163	27	(±40)
(G')	-480	163	53	(±60)
(G")	-187	8	-26	(±50)

L'examen de la troisième colonne montre que TOUTES les composantes Z sont NULLES - à la précision annoncée des observations.

En ce qui concerne (L), (G), (G'), (G"), ces faits indiquent que la vitesse absolue des différents groupes de galaxies est parallèle à (μ) .

On remarquera que ces observations mettent en évidence trois groupes de galaxies, (L), (G+G') et (G"), animés de vitesses très différentes; ceci se constate sur la figure 13, hodographe tracé dans le plan équatorial cosmique. Ces faits peuvent donc s'interpréter par l'existence d'une agitation générale des galaxies, tangentielle à leur stratification - et donc compatible avec la permanence de cette stratification.

L'interprétation de la vitesse du Soleil nécessite quelques remarques. La cinématique du Soleil est en effet composée de celle du centre de la Galaxie et de la vitesse orbitale du Soleil; comme le plan de la Galaxie est pratiquement perpendiculaire à (μ) (à 80°), il n'y a aucune raison a priori pour que cette dernière vitesse, de l'ordre de 200 km/s, soit parallèle à (μ). Ce qui se passe, c'est que le Soleil est actuellement proche du point "le plus haut" de sa trajectoire galactique (au maximum de Z): on le constate en remarquant la proximité dans le ciel du pôle cosmique (α) , dans Ophiuchus, et du centre galactique, dans le Sagittaire: ces points sont éloignés de 22° seulement.

L'interprétation finale est donc la suivante: la composante Z de la vitesse absolue du centre de la Galaxie est inférieure à 60 km/s.

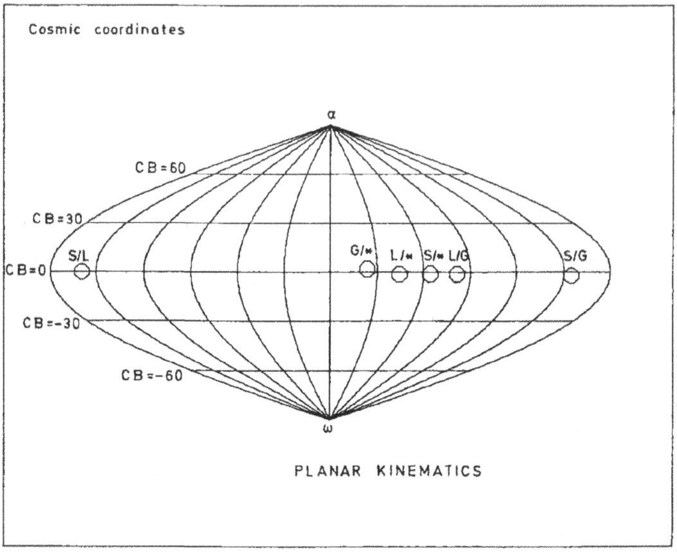

Figure 12

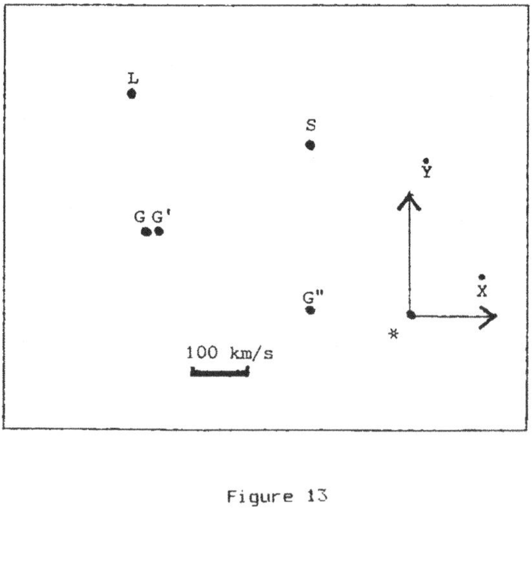

Figure 13

UN MODELE D'UNIVERS CONFRONTE AUX OBSERVATIONS

C O N C L U S I O N S

Nous venons de présenter une vue d'ensemble regroupant un grand nombre de faits, et tentant de leur donner une cohérence interne.

Si les observations confirment l'existence physique de la zone d'absence (μ), elle fournira une nouvelle voie d'accès à quelques questions fondamentales: genèse de la matière, dimensions, masse et évolution de l'Univers.

Il s'agit d'un modèle qui prend des risques: quelques observations contradictoires pourraient montrer que la zone d'absence (μ) n'est que coïncidence, ou tout au moins qu'elle n'a pas la netteté observée jusqu'ici.

Mais même dans cette éventualité, un certain nombre de faits indépendants restent acquis:

-Le modèle cosmologique de Friedmann - Lemaître obtenu colle à tous les faits observés - au moins aussi bien, dans chaque cas, que les autres modèles proposés: âge des étoiles et des galaxies, densité générale, statistique photométrique des galaxies et des quasars, etc.

-Fichtel et al. ont découvert un fond continu extra-galactique de rayons gamma, dans la zone 100-200 Mev; rayonnement dont l'origine constitue "a major open question". Or la répartition, le spectre et l'intensité de ce rayonnement sont précisément ceux que fait prévoir le présent modèle matière-antimatière, avec annihilation sur un "équateur" de l'espace.

-Les observations suivantes:

1) orientation dans le ciel du système entourant notre Galaxie, constitué de galaxies naines satellites et de nuages H I;
2) mêmes éléments pour la galaxie d'Andromède;
3) observations optiques et radio de M33;
4) répartition dans le ciel des galaxies du Groupe Local;
5) cinématique absolue de ce Groupe Local;
6) répartition dans le ciel des galaxies du Nuage Local;
7) parallélisme très marqué des halos de galaxies (galaxies diverses, échantillons de Rots et de Bosma);
8) répartition dans le ciel des galaxies du Super - Amas Local;
9) feuilletage de ce Super - Amas;
10) cinématique absolue des galaxies jusqu'à 30 Mpc.

sont chacune compatible avec une STRATIFICATION GENERALE DE NOTRE ENVIRONNEMENT, PARALLELE à la zone (μ).

REFERENCES

BOSMA, A.: 1981, Astron. J. 86, 1791
BURBIDGE, G.R., CROWNE, A.H., SMITH, H.E.: 1977, Astrophys. J. Suppl. 33, 113
CHENG, E.S., BOUGH, S., WILKINSON, D.T.: 1980, Bull. Amer. Astron. Soc. 12,488
CHENG, E.S., SAULSON, P.R., WILKINSON, D.T., COREY, B.E.: 1979, Astrophys. J. 232, L139
COREY, B.E.: 1978, Princeton University Dissertation
DE VAUCOULEURS, G.:1953, Astron. J. 58, 30
DE VAUCOULEURS, G.:1975, Astrophys. J. 202, 319
DE VAUCOULEURS, G., CORWIN Jr., H.G.: 1975, Astrophys. J. 202, 327
DE VAUCOULEURS, G., DE VAUCOULEURS, A., CORWIN, H.G.: 1976, "Second Reference Catalogue of Nearby Galaxies" University of Texas Press, Austin
DE VAUCOULEURS, G., PETERS, W.L.: 1981, Astrophys. J. 248, 395
DE VAUCOULEURS, G., PETERS, W.L., BOTTINELLI, L., GOUGUENHEIM, L., PATUREL, G.:1981, Astrophys. J. 248, 408
EMERSON,D.T., NEWTON,K.: 1978, I.A.U. Symposium 77 "Structure and properties of nearby galaxies", Eds. BERKHUIJSEN, E.M., WIELEBINSKI, R. p. 183
FICHTEL, C.E., SIMPSON, G.A., THOMPSON, D.J.: 1978, Astrophys. J. 222, 833
FLICHE, H.H.: 1981, Thèse, Université de Provence, CPT-81 P.1282, C.N.R.S. Marseille
FLICHE, H.H., SOURIAU, J.M.: 1979, Astron. Astrophys. 78, 87
FLICHE, H.H., SOURIAU, J.M., TRIAY, R.: 1980, Prétirage CPT-80 P.1196, C.N.R.S. Marseille
FLICHE, H.H., SOURIAU, J.M., TRIAY, R.: 1982 I, Astron. Astrophys. 108, 256
FLICHE, H.H., SOURIAU, J.M., TRIAY, R.: 1982 II, Prétirage CPT-82 P.1402, C.N.R.S.Marseille
FLICHE, H.H., SOURIAU, J.M., TRIAY, R.: 1982 III, en préparation
GORENSTEIN, M.V.: 1978, University of California Dissertation
GUNN, J.E.: 1978, "Observational Cosmology", Eds. MAEDERER, A., MARTINET, L., TAMMANN, G.A., Geneva Observatory
GUNN, J.E., OKE, J.: 1975, Astrophys. J., 195 255
GUNN, J.E., TINSLEY, B.M.: 1975, Nature 257, 454
HART, L., DAVIES, R.D.: 1982, Nature 297, 191
HOESSEL, J.G., GUNN, J.E., THUAN, T.X.: 1980, Astrophys. J., 241, 486
HUCHTMEIER, W.K.: 1978, I.A.U. Symposium no 77 "Structure and properties of nearby galaxies", Eds. BERKHUIJSEN, E.M., WIELEBINSKI, R. p. 197
JUNKKARINEN V.T., BURBIDGE E.M., SMITH H.E.: 1982, "PHL5200", U.C. San Diego preprint
LIFSHITZ, E.M., KHALATNIKOV, I.M.: 1963, Adv. Phys. 12, 185 (trad. anglaise)
MAC GILLIVRAY, H.T., DODD, R.J., MAC NALLY, B.V., CORWIN Jr, H.G.: 1982, MN 198, 605
OMNES, R.: 1979, "Cosmologie Physique", Les Houches, Eds. BALIAN, R., AUDOUZE, J., SCHRAMM, D.N., North Holland
OSMER, P.S.: 1982, "Pour La Science" 54, p. 98
OSTRIKER, J.P., PEEBLES, P.J.E.: 1973, Astrophys. J. 186, 467
PEEBLES, P.J.E.: 1979, Astrophys. J. 84, 730
PETERSON, B.A.: 1982, Communication Privée
ROTS, A.H.: 1980, Astron. Astrophys. Suppl. Ser. 41, 189
SANDAGE, A., HUMPHREYS, R.M.: 1980, Astrophys. J. 236, L1
SARGENT W.L.W., YOUNG P., BOKSENBERG, A. TYTLER, D. 1980: Astrophys. J. Suppl. 42, 41
SARGENT W.L.W., YOUNG P., SCHNEIDER D.P.: 1982, Astrophys. J. 256, 374
SCHATZMAN, E.: 1982, Cours de Goutelas (à paraître)
SMOOT, G.F., GORENSTEIN, M.V., MULLER, R.A.: 1977, Phys. Rev. Letters 39, 898
SOURIAU, J.M.: 1964, "Géométrie et Relativité" Hermann
SOURIAU, J.M.: 1974, Colloques Internationaux C.N.R.S. no 237, p 59
SOURIAU, J.M.: 1976, Journées Relativistes, Université Libre de Bruxelles
SOURIAU, J.M.: 1980, Proc. Colloque du centenaire d'Einstein, C.N.R.S., p. 197
STEIGMAN, G.: 1979, "Cosmologie Physique", Les Houches, Eds. BALIAN, R., AUDOUZE, J., SCHRAMM, D.N., North Holland
TAMMANN,G.A., SANDAGE,A., YAHIL,A.: 1979,"Cosmologie Physique",Les Houches, Eds. BALIAN, R., AUDOUZE, J., SCHRAMM, D.N., p. 56, North Holland
TRIAY, R.: 1981, Thèse de de 3e cycle, Université de Provence, CPT-81 P.1297, C.N.R.S. Marseille
TRIAY, R.: 1982, Prétirage CPT-82 P.1378, C.N.R.S. Marseille
TULLY, R.B.: 1982, Astrophys. J. 257, 389
WEINBERG, S.: 1978, "Les Trois Premières Minutes de l'Univers", Editions du Seuil
WOLF, J.A.: 1967, "Spaces of Constant Curvature", Mc Graw Hill

ALMOST MATHIEU EQUATION FOR SMALL AND BIG COUPLING CONSTANTS

by

D. TESTARD

Centre Universitaire d'Avignon
Faculté des Sciences
33, rue Louis Pasteur
84000 AVIGNON

and

Centre de Physique Théorique de Marseille
Case 907 - C.N.R.S. - Campus de Luminy
Route Léon Lachamp
13288 MARSEILLE Cédex 9
FRANCE

In this talk, I give a report of a joint word with J. BELLISSARD and R. LIMA about the spectrum of the Almost Mathieu hamiltonian [1]. This operator acting on $\ell^2(\mathbb{Z})$ is defined by :

$$\left(H(\theta,x,\mu)\,\psi\right)(n) = \psi(n+1) + \psi(n-1) + 2\mu\cos 2\pi(n\theta+x)\,\psi(n) \qquad (1)$$

where θ and x are elements of the one-dimensional torus $\mathbb{T}$ and μ is a real constant. The interesting case is for θ irrational.

This operator appears as an approximation of the hamiltonian for systems considered in solid state physics (one-electron approximation for a crystal with a magnetic field or tight-binding approximation for organic conductors with an irrational modulation due to the Peierls instability [2][3]).

Recently [3], G. ANDRE and J. AUBRY conjectured that the hamiltonian (1) exhibit a metal-insulator transition. Precisely, at least for almost all θ, one can expect that the operator $H(\theta,x,\mu)$ has an absolutly continuous spectrum for $|\mu|<1$ and, for $|\mu|>1$, a point spectrum with accumulating eigenvalues. My aim, is to give a rigourous statement in the direction of this conjecture. For others results about the Almost-Mathieu operator see [4][5].

In a first section we fix notations and recall essential facts about RÜSSMANN'S approximation functions [6]. In section two, we give precise statement of results. Indications on proofs are given in section three.

I - NOTATIONS

A RÜSSMANN's approximation function (RAF) is a strictly positive function Ω on $[0,\infty)$ which is continuous, decreasing, going to zero at infinity, such that $\frac{1}{s}\operatorname{Log}\frac{1}{\Omega(s)}$ is decreasing and $\int_{s_0}^{\infty}\frac{ds}{s^2}\operatorname{Log}\frac{1}{\Omega(s)}<\infty \quad (\forall s_0>0)$.

RAF's are usefull in order to give estimations on the loss of analyticity coming from small denominators problems [5]. For precise estimates, it is useful to introduce, for a given RAF Ω , the real functions $(\forall m\in\mathbb{N})\ (\forall\rho>0)$

$$\phi_m(\rho) = \int_0^{\infty}\frac{e^{-s}}{\Omega\left(\frac{s}{\rho}\right)}\,ds$$

$$\Psi_m(\rho) = \operatorname{Inf}\left\{\prod_{n=0}^{\infty}\phi_m(r_i)^{2^{-n-1}}\ \middle|\ 0<\dots\le r_n\le r_0,\ \sum_{n=0}^{\infty}r_n\le\rho\right\}$$

We also introduce notations for the function spaces of interest in the sequel. For a given set $\mathcal{K}$ in $\mathbb{T}+i\mathbb{R}=\mathbb{C}/\mathbb{Z}$, we will consider the set of Lipschitz functions $\mathcal{L}(\mathcal{K},\mathcal{B})$ from $\mathcal{K}$ to a Banach space $\mathcal{B}$. In practice $\mathcal{B}$ will be a Banach algebra of $\mathbb{C}$ or $M_2(\mathbb{C})$ valued functions defined on a analytic manifold with boundary $\mathcal{M}$, continuous, bounded on $\mathcal{M}$ and holomorphic in the interior of $\mathcal{M}$. We use the corresponding notation $\mathcal{L}(\mathcal{K},\mathcal{M},\mathbb{C})$ or $\mathcal{L}(\mathcal{K},\mathcal{M},M_2(\mathbb{C}))$ for these sets of functions. Specifically $\mathcal{M}$ will be a one-point set or

i) $\mathbb{T}_r = \mathbb{T}+i[-r,r] \qquad (r>0)$

ii) $\mathbb{D}_\lambda = \{z\in\mathbb{C}\mid |z|\le\lambda\} \qquad (\lambda>0)$

iii) $\mathbb{T}_r\times\mathbb{D}_\lambda \qquad (r,\lambda>0)$

Estimates of functions in $\mathcal{L}(\mathcal{K},\mathcal{B})$ will be obtained with respect the usual Lipschitz norm

$$\|F\| = \sup_{\phi\in\mathcal{K}}\|F(\phi)\|_{\mathcal{B}} + \sup_{\substack{\phi,\phi'\in\mathcal{K}\\ \phi\ne\phi'}}\frac{\|F(\phi)-F(\phi')\|_{\mathcal{B}}}{\|\phi-\phi'\|}$$

where $\|\ \|_{\mathcal{B}}$ is the norm in $\mathcal{B}$ and for $\phi\in\mathbb{T}+i\mathbb{R}$:

$$\|\phi\| = \inf_{n \in \mathbb{Z}} |\phi + n|$$

In examples given for $\mathcal{B}$, $\|\cdot\|_{\mathcal{B}}$ will be the supremum norm on $\mathcal{M}$. The corresponding Lipschitz norm will be denoted by $\|\cdot\|_r$, $\|\cdot\|_\lambda$, $\|\cdot\|_{r,\lambda}$ in cases i) ii) iii) respectively.

II - STATEMENT OF RESULTS

The main theorem is the following :

Theorem 1 : Let V be a complex valued continuous and bounded function on $\mathbb{T}_r (r>0)$, holomorphic in the interior of $\mathbb{T}_r$ and let

$$A = \sup_{|\operatorname{Im} y| \leq r} |V(y)|$$

Let $\theta \in \mathbb{T}$ satisfying

$$\|n\theta\| \geq \Omega(|n|)$$

for any $n \neq 0$, $n \in \mathbb{Z}$ and some RÜSSMANN approximation function Ω.

Let $\mathcal{K}$ be the subset of $\phi \in \mathbb{T}_R$ $(R>0)$ satisfying $(\forall n \in \mathbb{Z})$:

$$\|\phi + n\theta\| \geq \Omega(|n|)$$

i) Then there exists a constant M, such that if λ, ρ are such that

$$0 < \rho < r$$

$$0 < \lambda$$

$$A \lambda \Psi_2(2\pi\rho) < M$$

one can find $G \in \mathcal{L}(\mathcal{K}, \mathbb{T}_{r-\rho} \times \mathbb{D}_\lambda, M_2(\mathbb{C}))$ and $\alpha \in \mathcal{L}(\mathcal{K}, \mathbb{D}_\lambda, \mathbb{C})$ such that the following "twisted conjugacy" equation holds : $(\forall |\mu| < \lambda$, $\forall \phi \in \mathcal{K}$, $\forall y \in \mathbb{T}_{r-\rho})$

$$\begin{bmatrix} z+\mu V(y)+\alpha(\phi,\mu) & -1 \\ 1 & 0 \end{bmatrix} = G(\phi, y+\theta, \mu) \begin{bmatrix} z & -1 \\ 1 & 0 \end{bmatrix} G(\phi,y,\mu)^{-1}$$

where :

$$z = 2\cos\pi\phi \quad .$$

ii) G and α satisfy :

$$\det G(\phi,y,\mu) = 1$$

$$\alpha(\phi,0) = 0$$

$$G(\phi,y,0) = \mathbb{I}$$

Morever if $V(y)$ is real for real y , if μ is real, then $G(\bar\phi,y,\mu)$ and $\alpha(\bar\phi,\mu)$ are the complex conjugates of $G(\phi,y,\mu)$ and $\alpha(\phi,\mu)$ respectively.

The spectral properties of the Almost-Mathieu equation are the content of the following two results.

Theorem 2 : Let $V, \theta, \Omega, \mathcal{K}$ be as in theorem 1 and let $\mathcal{K}_0$ be the real part of $\mathcal{K}$: $\mathcal{K}_0 = \mathcal{K} \cap \mathbb{T}$. Then, there exists $\lambda_1 > 0$ such that for $|\mu| \leq \lambda_1$, the mapping

$$E^{\mu}: \quad \phi \longrightarrow E^{\mu}(\phi) = 2\cos\pi\phi + \alpha(\phi,\mu)$$

is an homeomorphism on $\mathcal{K}_0$. $E^{\mu}(\mathcal{K}_0)$ is a closed, non zero measure set if Ω satisfies :

$$\sum_{n=0}^{\infty} \Omega(n) < \frac{1}{4}$$

$E^{\mu}(\mathcal{K}_0)$ is included in the absolutely continuous part of the spectrum of the hamiltonian

$$\left(\mathrm{H}(\theta,x,\mu,V)\psi\right)(n) = \psi(n+1) + \psi(n-1) + \mu V(x+n\theta)\psi(n) \qquad (2)$$

acting on $\ell^2(\mathbb{Z})$, for any $x \in \mathbb{T}$.

Taking now, specifically $V(x) = 2\cos 2\pi x$, we have.

Theorem 3 : Assume θ to be Roth number (i.e : $\forall \varepsilon > 0$, there exists $C_\varepsilon > 0$ such that for any $p, q \in \mathbb{N}$

$$\left| \theta - \frac{p}{q} \right| > \frac{C_\varepsilon}{q^{2+\varepsilon}} \qquad).$$

Assume also that θ is not of constant type (i.e : the continued fraction expansion of θ

$$\theta = [a_1 [a_2 [a_3 [\cdots$$

is such that

$$\overline{\lim}\ a_i = \infty \qquad).$$

Given $\varepsilon > 0$, there exists $\lambda_2 > 0$ such that for almost all x and any $|\mu| < \lambda_2$ (resp : $|\mu| > \frac{1}{\lambda_2}$) the absolutely continuous part of the spectrum of $H(\theta, x, \mu)$ (1) (resp: the closure of the set of the eigenvalues of $H(\theta, x, \mu)$) has a Lebesgue measure greater than $4 - \varepsilon$ (resp : $(4-\varepsilon)|\mu|$)

If one notices that $4(1+|\mu|)$ is a trivial upperbound for the Lebesgue measure of the spectrum of $H(\theta, x, \mu)$, then, theorem 3 appears as an asymptotic version of the AUBRY-ANDRE conjecture which says that the spectrum of $H(\theta, x, \mu)$ for almost-all irrational θ's and almost all x has total length $4\,|1-|\mu||$.

III - INDICATIONS ON PROOFS

The method of proof of Theorem 1 is classical along the line of [6] or [7]. The main step is to solve the so-called linearized equation : the problem is to find $W \in \mathcal{L}(\mathcal{K}, \mathbb{T}_{r-\rho}, M_2(\mathbb{C}))(r>\rho>0)$ and $A \in \mathcal{L}(\mathcal{K}, M_2(\mathbb{C}))$ such that

$$W(\phi, x+\theta)\, M(\phi) - M(\phi)\, W(\phi, x) = F(\phi, x) + G(\phi, x+\theta)^{-1} A \; G(\phi, x) \tag{3}$$

for given F, G in $\mathcal{L}(\mathcal{K}, \mathbb{T}_r, M_2(\mathbb{C}))$ and

$$M(\phi) = \begin{bmatrix} 2\cos\pi\phi & -1 \\ 1 & 0 \end{bmatrix}$$

with a special form for A namely

$$A = \begin{bmatrix} \alpha & \beta \\ 0 & 0 \end{bmatrix} \tag{4}$$

It turns out to be possible, if G is sufficiently close to $\mathbb{I}$, using Fourier expansion of W with respect to x with an arbitrary small loss of analyticity (expressed by the fact that r is replaced by $r-\rho\ (\rho>0)$ with estimates :

$$\|A\| \leq C_1 \|F\|_r$$

$$\|W\|_{r-\rho} \leq C_2\ \phi_2(2\pi\rho)\ \|F\|_r$$

for some positive constants C_1, C_2.

This allows to use the RÜSSMANN's artificial parameter method [6]. In essence, it is a recursion process starting at :

$$\mathcal{N}_0(\phi, y, \mu) = M(\phi) + \delta \mathcal{N}_0(\phi, y, \mu) = M(\phi) + \mu \begin{bmatrix} V(y) & 0 \\ 0 & 0 \end{bmatrix}$$

More generally, if $\mathcal{N}_k \in \mathcal{L}(\mathcal{K}, \mathbb{T}_{r_k} \times \mathcal{D}_{\lambda_k}, M_2(\mathbb{C}))$ is given and such that :

$$\mathcal{N}_k(\phi, y, \mu) = M(\phi) + \delta \mathcal{N}_k(\phi, y, \mu) + G_k(\phi, y+\theta, \mu)^{-1} Z_k(\phi, \mu)\ G_k(\phi, y, \mu) \tag{5}$$

and $\delta \mathcal{N}_k(\phi, y, \mu)$ is of order μ^{2^k}, then solving (2) with $F = \delta\mathcal{N}_k$, $G = G_k$ in order to obtain W and A in $\mathcal{L}(\mathcal{K}, \mathbb{T}_{r_{k+1}} \times \mathcal{D}_{\lambda_{k+1}}, M_2(\mathbb{C}))$ it is possible to show that :

$$N_{k+1}(\phi, y, \mu) = \left(\mathbb{I} + W(\phi, y+\theta, \mu)\right)^{-1} N_k \left(\mathbb{I} + W(\phi, y, \mu)\right)$$

satisfies the equation (5) with k replaced by $k+1$, $\delta N_{k+1}(\phi, y, \mu)$ of order $\mu^{2^{k+1}}$

$r_{k+1} < r_k$, $\lambda_{k+1} \leq \lambda_k$, $Z_{k+1} = Z_k - A$ and

$$G_{k+1}(\phi, y, \mu) = G_k(\phi, y, \mu)\left(\mathbb{I} + W(\phi, y, \mu)\right)$$

By a specific choice of the sequences r_k, λ_k with λ_k strictly decreasing, it is possible to make all the process convergent :

$$r_k \to r - \rho > 0, \quad \lambda_k \to \lambda_\infty > 0, \quad G_k \to G_\infty, \quad N_k \to N_\infty, \quad \delta N_k \to 0,$$

$Z_k \to Z_\infty$ uniformly in the corresponding variables. This implies

$$N_\infty(\phi, y, \mu) = G_\infty(\phi, y+\theta, \mu)^{-1} N_0(\phi, y, \mu)\, G_\infty(\phi, y, \mu)$$

$$N_\infty(\phi, y, \mu) = M(\phi) + G_\infty(\phi, y+\theta, \mu)^{-1} Z_\infty(\phi, \mu)\, G_\infty(\phi, y, \mu)$$

which gives :

$$N_0(\phi, y, \mu) + Z_\infty(\phi, \mu) = G_\infty(\phi, y+\theta, \mu)\, M(\phi)\, G_\infty(\phi, y, \mu) \qquad (6)$$

By construction Z_∞ is of the form (4). Taking the determinant of both sides of (6) , one easily sees that $\beta \; (= \beta(\phi, \mu))$ is actually 0 . This is the content of theorem 1.

Before going to the proof of theorem 2, let us see why theorem 1 helps to solve the eigenvalue problem for the hamiltonian (2). (2) is a second order, finite difference operator and one easily sees that, to find a generalized eigenvector $(\psi_n)_{n\in\mathbb{Z}}$ with eigenvalue E for the hamiltonian (2) is equivalent to solve for a sequence of vectors $\Phi(n)$ in $\mathbb{C}^2$ the matrix equation :

$$\begin{bmatrix} E - \mu\left(V(x+n\theta)\right) & -1 \\ 1 & 0 \end{bmatrix} \Phi(n) = \Phi(n+1) \qquad (7)$$

By the twisted conjugacy of theorem 1, for small enough μ , this problem is now equivalent to solve the case where $\mu = 0$. It is well known that this is possible with E of the form $2\cos\pi\phi$ with corresponding $\Phi_0(n) = \begin{vmatrix} e^{\pm i\pi n\phi} \\ e^{\pm i\pi(n+1)\phi} \end{vmatrix}$

Using these facts one has that (7) can be solved with:

$$\Phi^{\pm}(n) = G_{\infty}(\phi, x+n\theta, \mu)\, \Phi_0^{\pm}(n)$$

and consequently the first component of $\Phi^{+}(n)$ (resp : $\Phi^{-}(n)$) are the components of a generalized eigenvector $\psi^{+}(n)$ (resp : $\psi^{-}(n)$) corresponding to the eigenvalue

$$E^{\mu}(\phi) = 2 \cos \pi \phi + \alpha(\phi, \mu)$$

and which is exponentially decreasing at $n \to +\infty$ (resp : $n \to -\infty$) if $\mathrm{Im}\, \phi > 0$.

Now, in order to prove theorem 2, it is sufficient to consider the classical formula for the resolvent of the operator (2)

$$\left(\left\{ \mathrm{H}(\theta, x, \mu, V) - E^{\mu}(\phi) \right\}^{-1} \chi \right)(m) =$$

$$\frac{1}{2i \sin \pi \phi} \left(\psi^{+}(m) \sum_{p=-\infty}^{m-1} \psi^{-}(p)\, \chi(p) + \psi^{-}(m) \sum_{p=m}^{\infty} \psi^{+}(p)\, \chi(p) \right)$$

and to get a uniform bound for its imaginary part, when $\mathrm{Im}\, \phi \to 0$. The crutial point is that for μ sufficiently small, $E^{\mu}(\phi) - E^{\mu}(\mathrm{Re}\, \phi)$ is not far from some multiple of $i\, \mathrm{Im}\, \phi$ due to the Lipschitz character of α with respect to ϕ .

Moreover, the same property allows to show that E^{μ} restricted to $\mathcal{K}_0$ is an homeomorphism and gives from a positive measure set an image with positive measure.

The proof of theorem 3 is obtained using two main informations.

The first one is obtained by using a coding of $\mathbb{T}$ which localizes points of the set $\mathcal{K}_0$ when Ω is restricted to satisfy conditions of the type :

$$\begin{aligned} &i) \quad \Omega(0) = \Omega(1) \\ &ii) \quad \sum_{n=1}^{\infty} \Omega(nq) \leq \delta \| q\theta \| \end{aligned} \tag{8}$$

($\forall q \in \mathbb{N}$ and for some $\delta > 0$). It is easily seen that, since θ is a Roth number, such a Ω exists. The main step is the following.

Proposition 4 : Let p_n/q_n be the successive convergents of the continued fraction of θ. Let $\mathcal{C}_n$ be the class of intervals of the form :

$$[j\theta,\ j\theta + q_n\theta - p_n] \quad (\bmod\ 1) \qquad 0 \leq j < q_{n+1}$$

which have an empty intersection with the set

$$A_{q_n} = \{ \phi \in \mathbb{T} \,/\, \|\phi - \mu\theta\| \leq \Omega(|\mu|) \quad \forall\, |\mu| \leq q_n \}.$$

then, for any $I \in \mathcal{C}_n$:

$$|I \cap \mathcal{K}_0| \geq (1 - 6\delta)\,|I|$$

where $|X|$ is the Lebesgue measure of the set X.

The proof of this proposition is an estimate of the Lebesgue measure of $I \cap \mathcal{K}_0^c$ $(I \in \mathcal{C}_n)$ using (8), the fact that if $I = [j\theta, j\theta + q_n\theta - p_n]$ intersects A_{q_n} then $j > q_n$ by definition of the class $\mathcal{C}_n$ and the classical property of successive convergent of :

$$q \neq 0\ ,\ \|q\theta\| \leq \|q_n\theta\| \quad \Rightarrow \quad q \geq q_{n+1}$$

which essentially says that the return time for the action by θ on $\mathbb{T}$ in an interval I of $\mathcal{C}_n$ is at least q_{n+1}.

It follows, from the proposition, using the individual ergodic theorem [8] that for almost all $\phi \in \mathbb{T}$, any n and any $I \in \mathcal{C}_n$ the orbit of ϕ by the irrational rotation $\phi \rightarrow \phi + 2\theta$ intersects $I \cap \mathcal{K}_0$. Taking n sufficiently big, one can even prove that the image by $\phi \longrightarrow 2\cos\pi\phi$ of the closure of $(\phi + 2\mathbb{Z}\theta) \cap \mathcal{K}_0$ has Lebesgue measure greater than $4 - \varepsilon/2$ for δ sufficiently small, using mutual disjointness of the interior of intervals in $\mathcal{C}_n$, which is a classical result in number theory. Using the fact that for μ sufficiently small, the Lipschitz norm of $\alpha(\cdot, \mu)$ is arbitrary small, one obtains the estimate

$$\left| E^{\mu}\left(\overline{\mathcal{K}_0 \cap (\phi + 2\mathbb{Z}\theta)}\right) \right| > 4 - \varepsilon$$

for small μ, and almost all ϕ.

The second information is connected to big μ and is known has the ANDRE-AUBRY duality. From the analysis in the proof of theorem 2, we know that for $\phi \in \mathcal{K}_0$, we have generalized eigenvectors of (1) with eigenvalues $E^{\mu}(\phi)$ which are of the form :

$$\psi(n) = f(x+n\theta)\, e^{i\pi n\phi}$$

where f is the restriction to $\mathbb{T}$ of an analytic function on $\mathbb{T}+i[r', r']$, $r' = r - \rho > 0$. Using the Fourier expansion of f :

$$f(x) = \sum_{p\in\mathbb{Z}} \tilde{f}(p)\, e^{2i\pi p x}$$

an easy computation shows that the $\tilde{f}(p)$'s must satisfy :

$$\tilde{f}(p+1) + \tilde{f}(p-1) + \frac{2}{\mu}\cos 2\pi\left(\frac{\phi}{2} + p\theta\right)\tilde{f}(p) = \frac{E^{\mu}(\phi)}{\mu}\tilde{f}(p)$$

Noticing that $H(\theta, x, \mu)$ and $H(\theta, x+n\theta, \mu)$ are conjugate, it follows that $\frac{1}{\mu}E^{\mu}(K_0 \cap (\phi+2\mathbb{Z}\theta))$ are eigenvalues of $H(\theta, \frac{\phi}{2}, \frac{1}{\mu})$ because analyticity of f entails exponential decrease for $\tilde{f}(p)$. This achieves the proof of theorem 3.

REFERENCES

1 J. BELLISSARD, R. LIMA, D. TESTARD.
About the spectrum of the Almost-Mathieu hamiltonian. Marseille Preprint (1982)

2 R.S. PEIERLS. Z. Phys 80 763 (1933)

3 G. ANDRE, S. AUBRY.
Analyticity Breaking and Anderson localization in Incommensurate lattices.
Ann. Israel Phys. Soc. 3 133 (1980)

4 A. Ya. GORDON. Usp. Math.Nauk. 31 257 (1976)

5 J. AVRON, B. SIMON .
Singular continuous spectrum for a class of almost-periodic Jacobi matrices.
(Princeton Uni. Preprint) (1980)

6 H. RÜSSMANN .
Ann. of New-York Acad. Scien. 387 90 (1980)

7 E. DINABURG, Ya SINAI
Functionnal Analysis Appl. 9 279 (1975)

8 P.R. HALMOS.
Lectures in Ergodic Theory. Chelsea. Pub. Company. New-York (1961)

PERIODIC SOLUTIONS OF HAMILTONIAN EQUATIONS

E. Zehnder
Mathematisches Institut der
Ruhr-Universität Bochum
4630 Bochum
(West-Germany)

Introduction

The search for periodic solutions in Hamiltonian systems is old and originated in frictionless mechanical problems such as the many body problem of celestial mechanics. It gave rise to many new ideas, tools and techniques useful in different branches of mathematics. For example, Poincaré, invastigating the restricted three body problem, was led to his continuation method. He also developped topological tools and formulated fixed point theorems for symplectic maps. In the twenties a very special problem from geometry, namely the problem of closed geodesics, which are periodic solutions of the geodesic flow, gave rise to two powerfull theories, namely the Morse-theory and the critical point theory of Ljusternik-Schnirelman, see R. Bott lectures [26] and [27]. These theories had been further developped and extended. In our connection we should mention the refinements of minimax techniques by P. Rabinowitz. The Morse-theory was generalized to continuous semiflows on metric spaces by C. Conley [40]. We shall actually describe this extension later on. We do however not follow up the geodesics problem and refer instead to Klingenbergs monography [59]. For more recent results we point out V. Bangert [14], V. Bangert and W. Klingenberg [15] and W. Ballmann, G. Thorbergsson, W. Ziller [12] and [13].

We are concerned with the general Hamiltonian equation on $\mathbb{R}^{2n}$:

$$\dot{x} = J\nabla h(x), \qquad x \in R^{2n},$$

where h is the Hamiltonian function defined on $\mathbb{R}^{2n}$, and where J is the standard symplectic structure on R^{2n}. We are looking for periodic solutions x(t) of the equation, i.e. solutions satisfying x(o) = x(T) for some T > o. There is a renewed interest in the existence problem of periodic solutions which is partly due to the fact that one is dealing with a challenging variational problem for a functional which is neither bounded from below nor from above. Also in the relations between Quantum mechanics and classical mechanics the periodic solutions of the corresponding classical system play a role, see S. Albeverio, P. Blanchard and R. Höegh-Krohn [106], we mention also J. Ralston [100] and Y. Colin de Verdière [101].

Our aim is to collect, in a rather unsytematic way, some old and more recent information. We sometimes indicate the ideas of the proofs of the statements in order to illustrate the variety of ideas and techniques these existence problems require.

Of course, periodic solutions are only a minor aspect in the understanding of the very complicated orbit structure of the flow of a Hamiltonian system. For special systems, namely the integrable equations and for systems close to these, great progress has been made in the K.A.M. theory. For a recent account of this theory we refer to the Doctoral dissertation by J. Pöschel [65]. According to this theory a big part of the phase space of such systems consists of quasiperiodic solutions with n rationally independent frequencies. We shall show later on, that the quasiperiodic solutions so found ly in the closure of the set of periodic solutions. These periodic solutions are, however, not of great interest as their periods are very large.

The organisation of this paper, which is an extended version of the talk given at the Bielefeld Encounters in Physics and Mathematics III (1982) is as follows. We first describe some local results in a neighborhood of an equilibrium point. Then a few global results on a given energy surface are described. The third part deals with forced oscillation problems for a Hamiltonian system which depends periodically on time.

Contents:

III. Forced oscillations

1. General results and a Morse-theory for periodic solutions
2. Hamiltonian vectorfields on a torus.

I. Periodic solutions locally in a neighborhood of an equilibrium point

1. Conditions at the linearized system

We start with the problem of finding periodic solutions of a equation $\dot{x} = f(x)$, $x \in R^m$, $f \in C^1(R^m)$ in a neighborhood of an equilibrium point, which we assume to be $x = 0$, so that $f(o) = o$. Let

$$\frac{\partial f}{\partial x}(o) = A \in L(R^m),$$

so that $\dot{x} = Ax$ is the linearized system. Clearly the presence of purely imaginary eigenvalues of A is necessary in order to find periodic solutions in a small neighborhood of $x = o$. In fact, if A is non-singular with none of its eigenvalues purely imaginary, then the rest point $x = o$ is a hyperbolic one and the flow of the vectorfield is, close to o, topologically conjugate to the flow of the linearized equation, which does not admit a periodic solution, expect the trivial one. If now A has the eigenvalues $\alpha_1, \ldots, \alpha_m$, not necessarily distinct, we shall assume, that

$$\alpha_1 = i\omega, \qquad \alpha_2 = -i\omega, \qquad \omega > o,$$

is a pair of purely imaginary eigenvalues with eigenvectors $A(e_1+ie_2) = i\omega(e_1+ie_2)$, so that $\dot{x} = Ax$ has the family $x(t) = \mathrm{Re}\{c(e_1+ie_2)e^{i\omega t}\}$ of periodic solutions having period $T = \frac{2\pi}{\omega}$, which fill out the plane $E = \mathrm{span}\{e_1, e_2\}$. We would like to find periodic solutions of the equation $\dot{x} = f(x)$ close to these. In general, such periodic solutions need not exist, as the following system in $\mathbb{R}^2$ shows:

$$\dot{x}_1 = -x_2 + r^2 x_1$$
$$\dot{x}_2 = x_1 + r^2 x_2 ,$$

where $r^2 = x_1^2 + x_2^2$. In fact, every solution $x(t)$ satisfies $\frac{d}{dt}|x(t)|^2 = 2|x(t)|^4$, so that for a $x(t)$ periodic with period τ we conclude $o = |x(\tau)|^2 - |x(o)|^2 = 2\int_0^\tau |x(t)|^4 dt$ and therefore $x(t) \equiv o$.

The class of vectorfields has therefore to be restricted. We shall assume that f has an integral $G \in C^2(R^m)$, i.e. a function satisfying $\frac{d}{dt} G(\phi^t(x)) = o$, if ϕ^t is the flow of f, or equivalently:

$$< \nabla G(x),\ f(x) > = o.$$

We furthermore assume that $G|E$ has nonvanishing Hessian at $x = o$. One verifies easily that $G_{xx}(o) = Q \in L(R^m)$ has then necessarily the form $< Q\xi, \xi > = \rho(\xi_1^2 + \xi_2^2)$, $\rho \neq o$, for $\xi = \xi_1 e_1 + \xi_2 e_2$, so that $d^2G|E$ is either positive or negative definite.

Theorem 1 (Lyapunov [60])

Let $\alpha_1 = i\omega$, $\alpha_2 = -i\omega$, $\omega > o$. *If the other eigenvalues of* A *satisfy*

$$\frac{\alpha_k}{\alpha_1} \neq \text{integer},\ 3 \leq k \leq m,$$

and if f *has a* C^2*-integral* G *with* $d^2G|E \neq o$, *say* $d^2G|E > o$, *then for every* ε *small there is a unique periodic solution* $x(t,\varepsilon)$ *near* E *having period* $T(\varepsilon)$ *near* $T = \frac{2\pi}{\omega}$ *and lying on the integral surface* $G(x) - G(o) = \varepsilon^2$. *Moreover* $x(t,\varepsilon) \to o$ *and* $T(\varepsilon) \to T$ *as* ε *tends to* o.

Postponing the easy proof we observe that the nonresonance condition $\frac{\alpha_k}{\alpha_1} \neq$ integer, $k \geq 3$ requires that the plane E contains all periodic solutions of $\dot{x} = Ax$ having period $T = \frac{2\pi}{\omega}$. The theorem applies in particular to a Hamiltonian vectorfield $f(x) = J\nabla H(x)$, $x \in R^{2n}$, which has the function $H \in C^2(R^{2n})$ as integral. We point out, that by an additional argument, it can be shown in case $f \in C^{r+1}$ and $G \in C^{r+1}$, $r \geq 1$, that the periodic solutions found fill out a 2-dimensional embedded C^r-manifold $M \subset \mathbb{R}^m$ which is tangent to the plane E at $x = o$. In case f and G are analytic, the embedding of M is even analytic as was shown by C.L. Siegel, [97]. J. Alexander and J. Yorke observed in [1] that the above Lyaponov-problem can be converted into a Hopf bifurcation problem, from which one gets the smoothnes of the embedding of M for free, we refer to the monography of H. Amann [82].

Proof: We shall apply Poincaré's continuation method. Stretching the variables $x = \varepsilon y$, $\varepsilon > o$, we introduce a perturbation parameter ε which allows to work on a fixed domain, $|y| \leq 1$ say. The vectorfield then becomes $\dot{y} = g(\varepsilon,y) = Ay + O_1(\varepsilon)$. The system has, for every $\varepsilon > o$ the integral $F(\varepsilon,y) := \varepsilon^{-2}(G(\varepsilon y) - G(o)) = \frac{1}{2} < Qy,y > + O_2(\varepsilon)$. We look for periodic solutions on $F(\varepsilon,y) = 1$ which corresponds to $G(x) - G(o) = \varepsilon^2$.

As reference solution we take for $\varepsilon = o$, $x(t,o) := \mathrm{Re}\{c(e_1 + ie_2)e^{i\omega t}\}$, where T is the period and where c is determined such that $F(o,x(t,o)) = 1$. The Floquet-multipliers for this solution for $\varepsilon = o$ are determined as the eigenvalues of $e^{T.A.}$, $T = \frac{2\pi}{\omega}$, i.e. $\lambda_k = e^{2\pi \frac{\alpha_k}{\alpha_1}}$, $k = 1,2,\ldots,m$. Thus by the nonresonance assumption $\lambda_1 = \lambda_2 = 1$, but $\lambda_k \neq 1$, $k \geq 3$, so that there are precisely two Floquetmultipliers equal to one. Consequently, the linearization of the transversal map of the flow, restricted to the integral surface F = 1 has no eigenvalue equal to 1 and we conclude, by the Implicit-Function Theorem, a fixed point for small ε, which gives rise to the required continued periodic solution $x(t,\varepsilon)$ on $F(\varepsilon,y) = 1$. •

If there are several pairs of purely imaginary eigenvalues the theorem yields also several families of periodic solutions with periods close to the normal modes, provided however, the corresponding nonresonance conditions are satisfied. If the nonresonance conditions are violated, no periodic solutions exept the equilibrium point need exist. An example is given by the following Hamiltonian system, with the complex notation $z_k = x_k + iy_k$:

$$H(z,\bar{z}) = \frac{1}{2}(|z_2|^2 - |z_1|^2) + (|z_1|^2 + |z_2|^2)\ \mathrm{Re}(z_1\ z_2)$$

For the corresponding Hamiltonian equations $\dot{x}_k = H_{y_k}$, $\dot{y}_k = -H_{y_k}$, or in complex notation $\dot{z}_k = -2i\,\frac{\partial H}{\partial \bar{z}}$ one computes readily that

$$-\frac{d}{dt}\ \mathrm{Im}(z_1 z_2) = 2(\mathrm{Re}(z_1 z_2))^2 + 2|z_1|^2|z_2|^2 + (|z_1|^2 + |z_2|^2)^2$$

$$\geq 4(\mathrm{Re}(z_1 z_2))^2 + (|z_1|^2 + |z_2|^2)^2.$$

Since the right hand side is positive for $(z_1,z_2) \neq (o,o)$ we conclude that $z_1 = z_2 = o$ is the only periodic solution of the system. In this example, which is due to J. Moser [90] the eigenvalues of the linearized system are $\pm i$ hence purely imaginary and the nonresonance condition is not satisfied. We emphasise that the Hessian of H at o is indefinite, and its signature is zero.

In contrast, a very remarkable theorem due to A. Weinstein [76] states, that if the Hamiltonian function on R^{2n} is definite at the equilibrium point, e.g. positive definite, then on every energy surface $H(z) - H(o) = \varepsilon^2 > o$ there are n solutions with periods close to the periods of the linearized systems. No nonresonance conditions are required, but instead $H_{xx}(o)$ is required to be definite. In order to formulate the more general version due to J. Moser [61] we consider a Hamiltonian equation

$$\dot{x} = J\nabla H(x), \qquad x \in R^{2n} ,$$

with $H \in C^2(R^{2n})$. The linearized equation at $x = o$ is $\dot{x} = Jh_{xx}(o)x = Ax$.

Theorem 2 (A. Weinstein, J. Moser)
Assume that $R^{2n} = E + F$, *where* E *and* F *are invariant subspaces under* A, *such that all solutions of* $\dot{x} = Ax$ *with* $x \in E$ *have period* $T > o$, *while none of the solutions in* $F \setminus \{o\}$ *have this period. Assume, moreover that* $H_{xx}(o)|E > o$. *Then for sufficiently small* $\varepsilon > o$ *every energy surface* $H(x) = H(o) + \varepsilon^2$ *contains at least* $\frac{1}{2}$ dim E *periodic solutions having periods close to* T.

We shall sketch the proof. $A = \text{diag}(B,C)$ on $E + F$, and we may assume that $T = 1$ so that $e^B = 1$ on E and $\det(e^C - 1) \neq o$. Normalizing the period to 1 we are looking for a 1-periodic solution $x(o) = x(1)$ of

$$\dot{x} = \lambda\, J\nabla H(x) \quad \text{and} \quad H(x) = \varepsilon^2 .$$

A solution is a critical point of the following functional f which is defined on the linear space of 1-periodic functions $x(t) \in R^{2n}$:

$$f(x,\lambda) := \int_o^1 \{\tfrac{1}{2} < \dot{x}, Jx > - \lambda(H(x) - \varepsilon^2)\}dt$$

Instead of looking for critical points of this functional we reduce the problem to find critical points of a related functional, which is defined on a finite dimensional manifold. (Later on a different, Lyapunov-Schmidt reduction will be carried out in detail). Observe, that if $u : E \to \mathbb{R}^{2n} = E + F$ is a map, which carries the linear vectorfield B on E into the given vectorfield $\lambda\, J\nabla H$ on $u(E)$, i.e. $du(\xi)\, B\xi = \lambda\, J\nabla H \circ u(\xi)$, $\xi \in E$, then $u(e^{Bs}\xi)$ is a periodic solution of $\dot{x} = \lambda\, J\nabla H(x)$. This idea does not quite work, but it can be proved that there are, in a pointed neighborhood of $0 \in E$ a function λ close to 1, a map $u : E \to E + F$ close to the injection map and a vectorfield $v(\xi) = v_E(\xi)$ close to zero solving

$$du(\xi)\ B\xi = \lambda\ J\nabla H \circ u(\xi) - v(\xi).$$

Moreover $\lambda(\xi) = \lambda(e^{Bs}\xi)$ and $e^{Bs}v_E(\xi) = v_E(e^{Bs}\xi)$. In addition the following normalization conditions hold:

$$\int_o^1 e^{-Bs}u_E(e^{Bs}\xi)ds = \xi \text{ and } < v(\xi), B\xi > = o.$$

It is clear that if $v(\xi^*) = o$ then $u(e^{Bs}\xi^*)$ is a periodic solution of the Hamiltonian equation, and it remains to find zeroes of the vector-field v. Define the action functional S by

$$S(\xi) = \int_o^1 \{\frac{1}{2} < \frac{d}{dt}\ u, Ju > - \lambda(H-\varepsilon^2) \circ u\}dt,$$

with $u(t,\xi) = u(e^{Bt}\xi)$ and define the averaged Hamiltonian H^* by:

$$H^*(\xi) = \int_o^1 H(u(e^{Bt}\xi))dt.$$

The functions S and H^* are invariant under $\xi \to E^{Bs}\xi$. Using the above normalizations one can show that

$$v(\xi) = J_E\ \nabla S(\xi) \quad \text{if} \quad H^*(\xi) = \varepsilon^2.$$

Set $M := \{\xi | H^*(\xi) = \varepsilon^2\}$ and define $S^* = S|M$. Then one shows that if $\nabla S^*(\xi) = o$ then also $\nabla S(\xi) = o$. It remains to find critical points of S^* on M, both of which are invariant under the S^1-action $\xi \to e^{Bs}\xi$. Since M is diffeomorphic to a sphere $S^{2\nu-1}$, where $2\nu = \dim E$, one concludes, by a topological argument, that S^* has at least ν orbits of critical points on M. •

We should mention that the method applied here inspired the abstract bifurcation results of M. Bottkol [28] about periodic orbits bifurcating off a submanifold on which all orbits are periodic. For an extension of these results we refer A. Weinstein [89].

There are many bifurcation results off an equilibrium point available in the resonance case and in the not definite case. They require, of course, conditions on the nonlinear part of the vector-field. The following general statement, due to E. Fadell and P. Rabinowitz [53] concludes in contrast to theorem 2 also periodic solutions in the not definite case. Assume there is a splitting $R^{2n} = E + F$ under the linearized A such that all the solutions in $E \setminus \{o\}$ have period $T > o$ and are nontrivial, while no solution in $F \setminus \{o\}$ is T-periodic. Let 2ν be the signature of $H_{xx}(o)|E$. Then either every neighborhood of $x = o$ contains a nontrivial T-periodic solution or there are $|\nu|$ distinct branches of nontrivial periodic solutions having periods close but not equal to T. A special case was obtained by S. Chow and J. Mallet-Paret [34]. More special results in the resonance case are due to H. Duistermaat [44] and [46] to D.S. Schmidt [74], to J.A. Sanders [73] to mention just a few.

2. Nonlinear conditions

So far, periodic solutions close to an equilibrium point were concluded from assumptions on the linearized vectorfield only. The periodic solutions found are the so called normal modes having periods close to the period of the linearized equation. If, in contrast, the Hamiltonian vectorfield is sufficiently smooth and nonlinear, then an abbundance of periodic solutions with large periods are found close to an equilibrium point. To formulate a statement we shall assume all the eigenvalues of the linearized $\dot{x} = Jh_{xx}(o)x$ to be purely imaginary and distinct from each other, so that $i\alpha_1,\dots,i\alpha_n,-i\alpha_1,\dots,-i\alpha_n$ are all the eigenvalues. We may assume, by means of a linear symplectic change of coordinates, that the quadratic part of the Hamiltonian is

$$H_2(x) = \sum_{k=1}^{n} \alpha_k I_k(x),$$

with $x = (p,q)$ and $I_k(x) = \frac{1}{2}(p_k^2 + q_k^2)$. Hence H_2 represents a system consisting of n decoupled harmonic oscillators having the frequencies α_j. The nonlinearity can be formulated, independently of the local coordinates, by menas of the so called Birkhoff-normal form. Excluding for some $\ell \geq 4$ the low order resonances by requiring that

$$< \alpha, j > \neq o, \quad j \in \mathbb{Z}^n, \quad o < |j| \leq \ell,$$

where $|j| = |j_1| + \dots + |j_n|$, it can be shown that there is an analytic and symplectic change of coordinates $x = \psi(p,q)$, so that in the new coordinates the Hamiltonian is in the following Birkhoff-normal-form:

(1) $$H \circ \psi(p,q) = P_\ell(J_1,\dots,J_n) + O_{\ell+1}(p,q), \text{ where}$$

$$P_\ell(J_1,\ldots,J_n) = <\alpha,J> + \frac{1}{2} <\beta J,J> + \ldots$$

is a polynomial of degree $[\frac{\ell}{2}]$ in $J_1,\ldots,J_n$; where $J_k(p,q) = \frac{1}{2}(p_k^2+q_k^2)$. The polynomial P_ℓ is a symplectic invariant. It describes as Hamiltonian function an integrable system with the n involutive integrals J_k, $1 \leq k \leq n$, which are linearly independent on $\{x \in R^{2n} | J_1 \cdot J_2, \ldots, J_k(x) \neq o\}$. In action and angle variables, $p_\nu + iq_\nu = \sqrt{2y_\nu}\, e^{\theta_\nu}$, this integrable system is, on $T^n \times \mathbb{R}^n_+$, given by the Hamiltonian function

$$P_\ell(y_1,\ldots,y_n),$$

which is independent of the angle-variables, so that the Hamiltonian equations become:

$$\dot{\theta} = \frac{\partial}{\partial y} P_\ell(y) \quad \text{and} \quad \dot{y} = -\frac{\partial}{\partial \theta} P_\ell(y) = o.$$

Hence every torus $T^n \times \{y\}$ is invariant under the flow of this integrable system and the restriction of the flow onto this torus is a linear Kronecker flow given by the n frequencies $\frac{\partial}{\partial y} P_\ell(y) = \omega$: $\theta \to \theta + t\omega$.

We now make the cucial assumption that the Hamiltonian system is nonlinear by requiring that det $\beta \neq o$. This means that the frequencies $\omega(y)$ depend on the amplitudes, they varie from torus to torus. In fact $y \to \frac{\partial}{\partial y} P_\ell(y)$ is locally open and, close to the equilibrium point, there is a dense set of tori on which the solution of the normal form part of (1) are all periodic.

In view of (1), the system is, close to the equilibrium point, approximated by an integrable system. It is well known, by the so called KAM theory, that under the assumption det $\beta \neq o$ many of the qualitative phenomena of the integrable system $P_\ell(J)$ are also present in the full system (1) if only one looks at a small neighborhood of the equilibrium point. As for periodic solutions one concludes, in particular, the following statement, see C. Conley and E. Zehnder [41], about periodic solutions close to an elliptic equilibrium point.

Theorem 3

Let $\ell \geq 4$ *and* $H \in C^\nu$ *with* $\nu \geq \max\{4n, \ell+1\}$ *and assume* $<\alpha, j> \neq o$ *for* $o < |j| \leq \ell$ *and* det $\beta \neq o$. *Set* $B_r = \{x \in R^{2n} | J_k(x) < r^2,\ 1 \leq k \leq n\}$ *and denote by* $P_r \subset B_r$ *the closure of the set of periodic solutions which are contained in* B_r. *Then* P_r *has positive Lebesgue measure; Moreover:*

$$m(B_r \setminus P_r) = O(r^{\lambda})\ m(P_r), \qquad \lambda = \frac{1}{2}(\ell-3).$$

We point out that under only finitely many conditions for the Taylor-coefficients of H at the equilibrium point, which formulate the required nonlinearity, it is concluded that the closure of the set of periodic solutions is even of positive measure, provided the Hamiltonian system is sufficiently smooth however. The periods of the solutions found by the above theorem are very large, so not very interesting. The excessive smoothnes assumption is required for the present proof. See on the other hand the recent beautiful results on the twist theorem by J. Mather [102] which apply for $n = 2$ and which do not require such excessive smoothnes assumptions. The above theorem, formulated however for symplectic maps in a neighborhood of an elliptic fixed point, has applications to closed geodesics. We do, however, not follow up this line.

Proof. By the so called K.A.M.-theory, it is known that a small neighborhood of the equilibrium point is differentiably foliated over a Cantor set in n-dimensional embedded invariant tori on which the flow is conjugate to a linear Kronecker flow with n rationally independent frequencies. We refer to J. Pöschel's dissertation [65] which extends earlier results and proves, in particular, that also in the differentiable case $m(E) = O(r^{\lambda})\ m(B_r)$, where $E \subset B_r$ is the set left out by these invariant tori in B_r. It remains therefore to show that these tori are contained in the closure of the set of periodic solutions. Singling out an invariant K.A.M. torus T_{ω} with frequencies $\omega = (\omega_1, \ldots, \omega_n)$ there is, see [95], an open and symplectic neighborhood $V = T^n \times U$, $U \subset R^n$ an open neighborhood of zero, so that on V the Hamiltonian is of the form

$$H(\xi,\eta) = c + \langle \omega,\eta \rangle + \frac{1}{2} \langle Q\eta,\eta \rangle + O(|\eta|^3),$$

with H a function on $T^n \times U$. The given torus T_{ω} is described by $T^n \times \{o\}$. Moreover $\det Q \neq o$. Therefore, the time 1 map of the flow ϕ^t satisfies the requirements of the Birkhoff-Lewis theorem, for which we refer to J. Moser [62]: It is exact symplectic and of the form:

$$\phi^1 : \quad \begin{aligned} \xi_1 &= \xi + \omega + Q\eta + O(|\eta|^2) \\ \eta_1 &= \eta \qquad\qquad\quad + O(|\eta|^3) \, . \end{aligned}$$

Consequently, every open neighborhood $V_{\nu} = T^n \times (\{|\eta| < \frac{1}{\nu}\} \cap U)$ contains a periodic orbit of ϕ^t. Let p_{ν} be a sequence of points, each of which is on a periodic orbit of ϕ^t contained in V_ν then $\eta(p_\nu) < \frac{1}{\nu}$ and there is, by compactnes, a subsequence with $p_\nu \to p_\infty \in V$, since $\eta(p_\infty) = o$ we find $p_\infty \in T_\omega$. If P is the closure of the set of periodic solutions, then $\phi^t(P) = P$. Also, $\phi^t(p_\infty) = (\xi(p_\infty) + t\omega, o)$, and since ω is rationally independent $cl\{\phi^t(p_\infty) | t \in \mathbb{R}\} = T^n \times \{o\} = T_\omega$. Hence $T_\omega \subset P$ as we wanted to prove. •

If n = 2, then more can be said about the character and distribution of the periodic orbits. We recall V. Arnold's sketch about the generic behaviour of the iterates of a measure preserving map in the plane in a small neighborhood of an elliptic, stable fixed point. In every neighborhood of this fixed point orbits of elliptic and hyperbolic periodic orbits show up in between the invariant Moser-curves. The hyperbolic orbits give rise to homoclinic points and therefore to unstable hyperbolic Bernoulli-systems in which the hyperbolic periodic points are dense. The same picture is repeated in every neighborhood of an elliptic periodic point for higher iterates. For details and for a proof we refer to [94], see also [98]. More global and surprising results for monotone twist maps have quite recently been obtained by J. Mather [102] and [103] who used new variational techniques, see also A. Katok [105].

The results mentioned so far involve small perturbations and we shall describe next more global results concerning periodic solutions having prescribed energy or prescribed period.

II. Global results

1. Prescribed energy:

If the Hamiltonian system $\dot{x} = J\nabla H(x)$, $x \in \mathbb{R}^{2n}$, which we also write as $\dot{x} = X_H(x)$, is time-independent, then the function $H \in C^2(R^{2n})$ is an integral of the Hamiltonian vectorfield X_H, i.e. $H(\phi^t(x)) = H(x)$, if ϕ^t is the flow. This is, of course, well known and follows simply from the fact, that J is skewsymmetric, $\frac{d}{dt} H(\phi^t(x)) = < \nabla H(\phi^t), \frac{d}{dt}\phi^t > = < \nabla H(\phi^t), J\nabla H(\phi^t) > = o$. Hence, introducing the energy surface

$$M := \{x \in R^{2n} | H(x) = c\} \text{ some } c \in R,$$

we have $\phi^t(M) = M$, and assuming M to be regular, i.e. $\nabla H(x) \neq o$, if $x \in M$, we conclude that X_H is tangent to the submanifold $M \in \mathbb{R}^{2n}$. One may ask whether the energy surface M carries a periodic solution for X_H. In order to reformulate the question we first observe that *the energy surface M together with the symplectic structure determine the vectorfield X_H on M up to a nonvanishing factor.* In fact, assume

$$M = \{x | H(x) = c\} = \{x | F(x) = c'\}$$

for two defining functions with $\nabla H(x) \neq o$ and $\nabla F(x) \neq o$ if $x \in M$. Then $\nabla F(x) = \lambda(x) \nabla H(x)$ with $\lambda(x) \neq o$ if $x \in M$ and consequently $X_F = \lambda X_H$ on M, with $\lambda \neq o$. It follows that X_F and X_H have, on M, the same orbits, although their parametization will be different in general. Indeed if ϕ^t is the flow of X_H and ψ^t is the flow of X_F, then we have, on M;

$$\psi^s(x) = \phi^t(x),$$

$$t = t(s,x), \text{ with } \frac{dt}{ds} = \lambda(\phi^t(x)) \text{ and } t(o,x) = o.$$

In particular, X_H and X_F have, on M, the same periodic orbits. Assume, for example, that X_H has the regular energy surface $M := \{x \in \mathbb{R}^{2n} \mid \frac{1}{2} |x|^2 = R > o\}$ then all the solutions of X_H on M are periodic. In fact the flow ψ^t of $F(x) := \frac{1}{2} |x|^2$ is simply $\psi^t(x) = e^{tJ}x = (\cos t)x + (\sin t)Jx$, which is periodic of period 2π.

We now look for conditions of the energy surface M which allows to conclude that every Hamiltonian vectorfield X_H having M as regular energy surface admits a period solution on M. The first, surprising result is due to P. Rabinowitz [69] and A. Weinstein [79].

Theorem 4

Let $M = \partial C$ be the C^2-boundary of a bounded and strictly convex domain $C \subset \mathbb{R}^{2n}$. Then every Hamiltonian vectorfield X_H, $H \in C^2(\mathbb{R}^{2n})$, with energy-surface $M = \{x \in R^{2n} | H(x) = c\}$ for some $c \in \mathbb{R}$, and with $\nabla H(x) \neq o$ on M, possesses at least one periodic solution on M.

Actually, P. Rabinowitz proved a much stronger result, assuming C to be starlike with respect to an interior point.

Proof: The proof given below is based on an amazing idea due to F. Clark [36], which allows the application of the most simple direct methods of the calculus of variations. For another proof, which is based on the mountain-pass Lemma, a special minimax technique, we refer to A. Ambrosetti [4].

The first step is to choose a convenient function H representing M. We may assume $o \in C$, and using the convexity, we define a function F by, $F(o) = o$ and $F(x) = \lambda$ if $\xi = \lambda^{-1}x \in \partial C = M$. Then $F(\rho x) = \rho F(x)$, $\rho \geq o$. Moreover $F \in C^2(R^{2n} \setminus \{o\})$ and $F_{xx}|TM > o$. Define now $H(x) = (F(x))^2$. Then H satisfies $M = \{x \in R^{2n} | H(x) = 1\}$ and

i) $H(o) = o$, $H \in C^2(\mathbb{R}^{2n} \setminus \{o\})$

ii) $H(\rho x) = \rho^2 H(x)$, $\rho \geq o$.

iii) $H_{xx}(x) > o$, $x \neq o$.

In view of the observation previous to the theorem it is sufficient to find periodic solutions for this very special Hamiltonian system on M. Normalizing the period to 2π, we look for 2π-periodic solutions of

$$\dot{x} = \lambda J\nabla H(x) \quad \text{on} \quad M$$

for some $\lambda \neq o$. These equations are the Euler-equation of the variational principle

$$\min_{x \in H^1(S^1)} \int_o^{2\pi} H(x(t))dt \quad \text{and} \quad \frac{1}{2}\int_o^{2\pi} < Jx, \dot{x} > = 1.$$

But observing that $\frac{1}{C}|x|^2 \leq H(x) \leq C|x|^2$, since H is homogeneous of degree 2, one sees that the infimum is not taken on. In fact, $|\dot{x}_n| \to +\infty$, for every minimizing sequence, where $| \ |$ stands for the $L^2[o,2\pi]$-norm. Clark's idea now is to choose a different, better variational principle for the Legendretransformation G of H, where in contrast to the Legendretransformation customarily used in mechanics, all the variables are transformed: Define, using the convexity of H,

$$G(y) = \max_{\xi} (< \xi, y > - H(\xi)) = < x, y > - H(x),$$

then $G(y) + H(x) = < x, y >$, where $y = \nabla H(x)$, or equivalently $x = \nabla G(y)$. G has the same properties as H above and we consider the following alternative constrained variational principle:

$$\min_{z \in A} \int_o^{2\pi} G(\dot{z})dt = \min_{z \in A} I(z), \text{ where}$$

$$A = \{z \in H^1(S^1) \mid \frac{1}{2\pi} \int_0^{2\pi} z(t) = 0 \quad \text{and} \quad \frac{1}{2} \int_0^{2\pi} < Jz, \dot{z} > dt = 1\}.$$

This functional has the right compactnes conditions, it is bounded from below, and the infimum is taken on. In fact if $z \in A$, then $|z| \leq |\dot{z}|$ and hence $2 = (Jz, \dot{z}) \leq |z|\ |\dot{z}| \leq |\dot{z}|^2$. Moreover, since G is homogeneous $\frac{1}{C}\ |y|^2 \leq G(y) \leq C|y|^2$ for some constant $C > 0$. It follows that $I(z) \geq \frac{1}{C}\ |\dot{z}|^2 \geq \frac{2}{C} > 0$, so that

$$\inf_{z \in A} I(z) = \mu > 0.$$

Take a minimizing sequence $\mu = \lim_{j\to\infty} I(z_j)$, with $z_j \in A$. Then $2 \leq |\dot{z}_j|^2 \leq M$ for some $M > 0$, therefore z_j is bounded in $H^1(S^1)$ and con-consequently there is a subsequence z_j such that $z_j \to z_*$ weakly in $H^1(S^1)$ and $z_j \to z_*$ uniformly in $C(S^1)$. Hence $z_* \in A$ and we show that z_* is a minimum. From the convexity of G we conclude $\int_0^{2\pi} G(\dot{z}_*)dt - \int_0^{2\pi} G(\dot{z}_j)dt \leq \int_0^{2\pi} < G_y(\dot{z}_*),\ \dot{z}_*-\dot{z}_j > dt \to 0$ as $j \to \infty$, hence $\mu \leq \int_0^{2\pi} G(\dot{z}_*)dt \leq \underline{\lim} \int_0^{2\pi} G(\dot{z}_j)dt = \mu$, so that indeed:

$$I(z_*) = \mu.$$

One verifies immediately that this z_* satisfies the Euler-equation:

$$\nabla G(\dot{z}_*) = \alpha\ Jz_* + \beta \text{ for } \alpha = \mu \neq 0.$$

Now taking the pointwise Legendre-transformation we find $\dot{z}_* = \nabla H(\alpha\ Jz_* + \beta)$ so that, in particular, $z_* \in C^2[0, 2\pi]$. Define now $x(t) = c(\alpha\ Jz_* + \beta)$, then, by the homogeneity of ∇H, $\dot{x}(t) = \alpha\, J\nabla H(x(t))$ and choosing the constant c such that $\int_0^{2\pi} H(x(t))\ dt = 2\pi$, we conclude that $y(t) := x(\alpha^{-1}t)$ is a solution of $\dot{x} = J\nabla H(x)$ on $H = 1$ having period $2\pi\mu > 0$, as we wanted to prove. •

The above trick, sometimes called the dual action principle and used often in optimization theory stimulated many results for Hamiltonian having certain convexity properties, see I. Ekeland [48-52]. It has also been applied in existence problems of semilinear partial differential equations, see H. Brezis and L. Nirenberg [29].

One may ask, whether there are more than just one periodic solution on M. Restricting the shape of the convex region C further, I. Ekeland and J. Lasry [52] succeeded in proving the following multiplicity result, which can be viewed as a global version of the normal-mode theorem of A. Weinstein above.

Theorem 5

Assume, in addition to the assumptions of the previous theorem, that $D_r \subset C \subset D_R$, *for two discs with radii* $r < R < \sqrt{2}\, r$. *Then* $M = \partial C$ *carries at least* n *distinct periodic orbits.*

The additional periodic orbits are found as critical points of a related functional by means of a minimax technique, which is based on an index theory involving an S^1-action. The quantitative restrictions on C make shure that the different critical points found correspond to geometrically different periodic orbits, a notorious difficulty in multiplicity theory which is well known from the problem of closed geodesics. The index theory which replaces analogous arguments based on category-theory is possible in the time-independent case. It was introduced by E. Fadell and P. Rabinowitz in [53], for another interpretation we refer to V. Benci [108]. The theory was used since then is several papers in order to establish multiplicity results for periodic orbits, so for instance in [3] and in [109]. We point out, that recently the proof of the above theorem was considerably simplified by A. Ambrosetti and G. Mancini [6]; there are also other proofs due to A. Hofer [96] and to M. Willem [81].

Let finally $H \in C^2(R^{2n})$ be in the special form of a classical conservative system in $x = (p,q) \in R^n \times R^n$ with a potential V:

$$H(x) = \frac{1}{2}|p|^2 + V(q).$$

Geometrical ideas inspired by H. Seifert [75] and related to the principle of least action of Euler-Maupertius-Jacobi led H. Gluck and W. Ziller [54] to the following result:

Theorem 6

Let $M = \{H(x) = c\}$ *be an regular energy surfaces. If* $\{q \in R^n | V(q) \leq c\}$ *is (non empty and) compact, then* M *carries a periodic orbit.*

Recalling the transformation properties of Hamiltonian vectorfields one concludes that every regular energy surface,which is symplectically diffeomorphic to a strictly convex one,carries a periodic solution. Such a situation is however very hard to recognize. Since the convexity property is generally lost under symplectic diffeomorphismus it would be interesting to find invariant conditions, for instance topological ones, on an energy surface guaranteeing a periodic solution; see also A. Weinstein [80] for symplectic conditions. In this connection one has to keep in mind that there are vectorfields on odd dimensional spheres having no periodic orbits, hence contradicting a Conjecture of Seifert, see P.A. Schweitzer [83] and T.W. Wilson [84]. Most likely such vectorfields do exist in the more restricted class of measure preserving vectorfields and one might conjecture that a regular energy surface which is diffeomorphic to a sphere admits one or several periodic orbits. It should be said that generically, in the C^1-sense, the periodic solutions of a Hamiltonian system are dense on a compact energy-surfaces. This is a consequence of C. Pugh and C. Robinson's closing Lemma [99].

2. Prescribed period

The periods of the periodic solutions found above are not known. Under specific conditions on H there are estimates of the periods in dependence on the "shape" of C available, see P. Rabinowitz [70], and B. Croke and A. Weinstein [88]. One can ask however for periodic solution having a prescribed period instead of a prescribed energy. Clearly, the more twisting nonlinearity one globally postulates, the more likely it is to find many periodic solutions of a given period. Having found such a periodic solution one can also worry whether its period is actually a minimal period. There are plenty of results about these questions reached by means of various variational techniques. We do not describe them in detail but refer to [5], [50], [51], [39]. As for the asymptotically linear case we mention the results [2], [3], [41], [109] which are of more topological nature.

III. Forced oscillations

1. General results:

Does the periodically forced Duffing equation:

$$\ddot{x} + \alpha x + \beta x^{2m+1} = \gamma x^n \sin(2\pi t),$$

with $n \leq 2m$, which is periodic in time of period 1, possess periodic solutions of period 1? If $\beta > o$, then in fact there are infinitely many. Indeed, consider, more generally, the second order equation on $\mathbb{R}$:

$$(*) \qquad \ddot{x} + f(t,x) = o,$$

then the following result is due to P. Hartman [107]. It was preceded by a theorem of H. Jacobowitz [58], who had to impose an additional restriction.

Theorem 7

Let $f(t,x) \in C^1(\mathbb{R}^2)$ *satisfy the three conditions*

(i) $f(t+1),x) = f(t,x)$

(ii) $f(t,o) = o$

(iii) $\frac{f(t,x)}{x} \to +\infty$ as $|x| \to \infty$ uniformly in t.

Then the differential equation (*) *has infinitely many periodic solutions of period* 1. *More precisely, there exists an integer* N_o *such that for every integer* $N \geq N_o$ *equation* (*) *has a solution of period* 1 *which has exactly* 2N *zeroes in* $[o,1)$. *If* n_o *is the number of zeroes of a nontrivial solution* $g(t)$ *of* $\ddot{y} + f(t,o)y = o$ *in* $[o,1)$ *then one can take* N_o *as the smallest integer* $\geq \frac{1}{2} n_o + 1$. *Also, for any two integers* p,q *with* $p \geq N_o q$ *there exists a periodic solution having period* q *and with precisely* 2p *zeroes in* $[o,q)$.

The nontrivial proof is an application of the global geometric Birkhoff fixed point theorem, after an a priori estimate for the solutions having a prescribed number of zeroes in $[o,1)$. We should point out right away that there is no genuine generalization of this theorem to higher dimensions, i.e. to the general, time-dependent Hamiltonian system

$$(2) \qquad \dot{x} = J\nabla H(t,x), \ (t,x) \in R \times \mathbb{R}^{2n},$$

$n \geq 2$, with $H \in C^2(R \times \mathbb{R}^{2n})$ being periodic in t:

$$H(t+T,x) = H(t,x)$$

for some $T > o$. We look for solutions $x(t)$ which are periodic having period T, i.e. $x(t+T) = x(t)$.

Under various assumptions which postulate a strong nonlinear behaviour of the system at infinity the existence of at least one T-periodic solution can be guaranteed, see P. Rabinowitz [68], I. Ekeland [47]. As for multiplicity results we mention the following statement by A. Bahri and H. Berestycki [86] about a system close to a highly nonlinear, but time-independent system:

Theorem 8

Assume $H(t,x) = H_o(x) + \langle f(t),x \rangle \in C^2(R \times R^{2n})$ *with* $f(t+T) = f(t)$. *Moreover, assume* $H(x) \leq \theta \cdot \langle \nabla H(x),x \rangle + C$ *for all* $x \in \mathbb{R}^{2n}$ *with some* $o < \theta < 1/2$, *and* $a|x|^{p+1} - b \leq H(x) \leq c|x|^{q+1} + d$ *for all* $x \in \mathbb{R}^{2n}$, *with* $1 < p \leq q < 2p+1$. *Then there are infinitely many* T-*periodic solutions.*

The intricate proof of this theorem, based on approximation and minimax techniques introduced by P. Rabinowitz makes use of the S^1-action of the time-independent problem. In contrast to this analytically difficult result we shall describe next some existence results of more topological nature for systems which are assumptotically linear.

We first recall that the Hamiltonian equations (2) are the Euler-equations of the following functional f, defined on the linear space of T-periodic functions $t \to x(t) = x(t+T) \in R^{2n}$:

$$(3) \qquad f(x) = \int_o^T \{\tfrac{1}{2} \langle \dot{x},Jx \rangle - H(t,x(t))\}dt.$$

The critical points of f are indeed the required periodic solutions:

$$f'(x)y = \frac{d}{d\varepsilon} f(x+\varepsilon y)\Big|_{\varepsilon=o} = \int_o^T \langle -J\dot{x} - \nabla H(t,x), y \rangle \, dt = (\nabla f(x),y).$$

Therefore $f'(x)y = o$ for all y, is equivalent to $\nabla f(x) = o$ i.e. to $\dot{x} = J\nabla H(t,x)$, since $J^2 = -1$. The difficulty in finding critical points of f stems from the fact, that f is neither bounded from below nor from above. In contrast to the closed geodesic problem, the critical points are saddlepoints infinite dimensional stable and infinite dimensional unstable invariant manifolds, which can be read of from the Hessian of f at a critical point x:

$$f''(x)(y_1,y_2) = \int_0^T < -J\dot{y}_1 - H''(t,x(t))y_1, y_2 > dt.$$

In order to find critical points of f we shall apply Morse-theory and study the gradient flow on the infinite dimensional loop space

$$\frac{d}{ds} x = \nabla f(x).$$

The equilibrium points of this gradient equation are then the required periodic solutions. In order to label the periodic solutions, we shall introduce next an index for a periodic orbit, which is similar to the Maslov index, for which we refer to [45].

Let $x_o(t) = x_o(t+T)$ a closed curve. Then we look at the linearized equation along $x_o(t)$:

$$\dot{y} = JH''(t,x_o(t))y =: JA(t)y,$$

with $A(t) = A(t+T)$. Let $X(t)$ be the fundamental solution satisfying $\dot{X}(t) = JA(t)\ X(t)$ and $X(o) = id$, then the eigenvalues $\sigma(X(T))$ are called the Floquet multipliers of $x_o(t)$.

Definition: A periodic solution $x_o(t)$ is called nondegenerate, if 1 is not a Floquetmultiplier of $x_o(t)$. The definition requires that the above linear system admits no nontrivial T-periodic solution, as is well known from Floquet-theory.

Now, the arc $X(t)$, $o \le t \le T$ is in $Sp(n,\mathbb{R})$ with $X(o) = 1$ and $X(T) \in W^* := \{M \in Sp(n,R) \mid 1 \notin \sigma(M)\}$, if $x_o(t)$ is nondegenerate. Let $M = PO$ be the unique polardecomposition, $P = (MM^T)^{1/2}$ being positive symmetric and symplectic, and O orthogonal and symplectic, hence

$$O = \begin{pmatrix} u_1 & -u_2 \\ u_2 & u_1 \end{pmatrix}, \quad \text{with} \quad \bar{u} := u_1 + iu_2 \in L(\mathbb{C}^n) \text{ unitary.}$$

If $\gamma(t) \in Sp(n,R)$ is an arc there is the associated arc $\bar{u}(t) \in U(n)$ of unitary matrices. We pick a continuous function $\Delta(t) \in \mathbb{R}$ with det u(t) = exp (i $\Delta(t)$). Then $\Delta(1)-\Delta(o) = \Delta(\gamma)$ depends only on γ, and if $\gamma(o) = \gamma(1)$ then $\Delta(\gamma) = 2\pi m$, $m \in \mathbb{Z}$. It can be shown that a loop γ is contractible in Sp(n,R) if and only if $\Delta(\gamma) = o$. Let now $\gamma(t) = X(t)$, $\sigma \leq t \leq T$, extend it to an arc $\bar{\gamma}$ by connecting X(T) within W* to either M_+, if the degree of X(T) is (+1) or to M_- if the degree of X(T) is (-1), where

$$M_+ = -1$$

$$M_- = \begin{pmatrix} 2 & & & 0 \\ & -I & & \\ & & \frac{1}{2} & \\ 0 & & & -I \end{pmatrix} .$$

Since $n \geq 2$, there are many logarithm's, namely $M = e^{JA_\ell^\pm}$, $\ell \in \mathbb{Z}$, where

$$A_\ell^+ = \left(\begin{array}{cccc|cccc} \pi & & & & & & & \\ & (2\ell+1)\pi & & & & 0 & & \\ & & \pi & & & & & \\ & & & \ddots & & & & \\ & & & \pi & & & & \\ \hline & & & & \pi & & & \\ & & & & & (2\ell+1)\pi & & \\ & 0 & & & & & \pi & \\ & & & & & & & \ddots \\ & & & & & & & \pi \end{array}\right)$$

$$A_\ell^- = \left(\begin{array}{cccc|cccc} 0 & & & & \ln 2 & & & \\ & (2\ell+1)\pi & & & & & & \\ & & \pi & & & 0 & & \\ & & & \ddots & & & & \\ & & & \pi & & & & \\ \hline \ln 2 & & & & 0 & & & \\ & & & & & (2\ell+1)\pi & & \\ & & & & & & \pi & \\ & 0 & & & & & & \ddots \\ & & & & & & & \pi \end{array}\right) .$$

Let now $\gamma_1 = \bar{\gamma} \cup (-\hat{\gamma}_\ell)$ with $\hat{\gamma}_\ell(t) = \exp(tJA_\ell)$ then $\Delta(\gamma_1) = \Delta(\bar{\gamma}) - \Delta(\hat{\gamma}_\ell) = 2\pi m$. Set $s = \ell + m$ and define the loop $\gamma_2 = \bar{\gamma} \cup (-\hat{\gamma}_s)$, so that $\Delta(\gamma_2) = o$ and γ_2 is contractibel in $Sp(n, \mathbb{R})$. In this case we define the index of the period orbit $x_o(t)$, or of the linear system $\dot{x} = JA(t)x$, with $A(t) = H''(t, x_o(t))$ to be

$$j(x_o(t)) = j(A(t)) = \begin{cases} 2s + (n-1) & \text{if } \deg X(T) = +1 \\ 2s + n & \text{if } \deg X(T) = -1. \end{cases}$$

The integer j does not depend on the extension in W^* of the arc to $M_\pm$. If P denotes the set of linear equations $\dot{y} = JA(t)y$, with $1 \notin \sigma(X(T))$, then it can be decomposed into the equivalence classes of equations in P which can continuously be deformed into each other within the class P. The above considerations show, that every equivalence class contains an equation $\dot{y} = JA_\ell^\pm y$ with constant coefficients, moreover the equivalence classes are characterized by the index $j = j(A_\ell^\pm)$. For details see [41].

After these explanations we formulate an existence statement for periodic solutions of an asymptotically linear Hamiltonian system, due to C. Conley and E. Zehnder [41]:

Theorem 9 (A Morse-theory statement)

Let $h = h(t,x) \in C^2(R \times R^{2n})$, $n \geq 2$, *be periodic in time of period* $T > o$, $h(t+T, x) = h(t,x)$. *Assume* (i) *the Hessian of* h *is bounded:* $-\alpha \leq h''(t,x) \leq \alpha$ *for all* $(t,x) \in R \times R^{2n}$ *and for some constant* $\alpha > o$. *Assume* (ii) *the Hamiltonian vectorfield to be asymptotically linear*

$$Jh'(t,x) = JA_\infty(t)x + o(|x|), \text{ as } |x| \to \infty$$

uniformly in t, *where* $A_\infty(t) = A_\infty(t+T)$ *is a continuous loop of symmetric matrices. Assume* (iii) *that the trivial solution of the equation* $\dot{x} = JA_\infty(t)x$ *is nondegenerate and denote its index by* j_∞. *Then the following statements hold:*

(1) *There exists a periodic solution of period* T *for* (2). *If this periodic solution is nondegenerate with index* j_o, *then there is a second* T*-periodic solution, provided* $j_o \neq j_\infty$. *Moreover if there are two nondegenerate periodic solutions there is also a third periodic solution.* (2) *Assume all the periodic solutions are nondegenerate, then there are only finitely many of them and their number is odd. If*

j_k, $1 \leq k \leq m$, *denote their indices we have the following identity:*

$$\sum_{k=1}^{m} t^{-j_k} = t^{-j_\infty} + t^{-d}(1+t)\, Q_d(t),$$

where $d > o$ *is an integer, and where* $Q_d(t)$ *is a polynomial having non-negative integer coefficients.*

The theorem extends earlier results in [2] and by K.C. Chang [31]. We point out an interesting special case of the above statement, which can be viewed as a generalization to higher dimensions of the Poincaré-Birkhoff fixed point theorem for mappings in the plane. This well known theorem states that a measure preserving homeomorphism of an annulus, which twists the two boundaries in opposite directions has at least two fixed points, see G.D. Birkhoff [24] and, more recently, M. Brown and W.D. Neumann [30], see also P.H. Carter [92] and A. Chenciner [93].

Corollary.

Let $h = h(t,x) \in C^2(R \times R^{2n})$, $n \geq 2$ *be periodic,* $h(t+T),x) = h(t,x)$ *and let the Hessian of* h *to be bounded. Assume*

$$Jh'(t,x) = JA_\infty(t)x + o(|x|) \text{ as } |x| \to \infty$$

$$Jh'(t,x) = JA_o(t)x + o(|x|) \text{ as } |x| \to o$$

uniformly in t, *for two continuous loops* $A_o(t+T) = A_o(t)$ *and* $A_\infty(t+T) = A_\infty(t)$. *Assume that the two linear systems* $\dot{x} = JA_\infty(t)x$ *and* $\dot{x} = JA_o(t)x$ *do not admit any nontrivial* T*-periodic solutions, and denote by* j_∞ *and* j_o *the indices of these two linear systems. If* $j_\infty \neq j_o$ *then there exists a nontrivial* T*-periodic solution of* (2). *Moreover, if this periodic solution is also nondegenerate then there is a second* T*-periodic solution.*

In other words, if the two linear system with $A_o(t)$ and $A_\infty(t)$ cannot be continuously deformed into each other within the set P, then we conclude the existence of a T-periodic orbit. The corollary only claims the existence of one T-periodic solution except if the nondegeneracy condition is satisfied. This is in contrast to the Poincaré-Birkhoff fixed point theorem which always guarantees two fixed points. Birkhoff's original proof in [23] also suggests, that the integer $|j_o - j_\infty|$ is a measure for the lower bound of the number of

periodic solutions of (2).
As a sideremark we observe, however, that under additional assumptions the following result has been proved in [3] by means of mini-max techniques:

Theorem 10

Let h *be as in the Corollary and assume, in addition,* $h(t,x) = h(t,-x)$ *for all* $(t,x) \in R \times R^{2n}$. *Moreover, let* $A_o(t) = A_o$ *and* $A_\infty(t) = A_\infty$ *be independent of* t. *Then* (1) *has at least* $|j_o - j_\infty|$ *nontrivial pairs* $(x(t), -x(t))$ *of* T-*periodic solutions.*

We sketch the proof of theorem 9. The required periodic solutions are found as the critical points of the functional f given by (3). To be precise we introduce the Hilbert space $H = L_2((o,1)) : R^{2n})$. Define in H the linear operator $A : dom(A) \subset H \to H$ by setting $dom(A) = \{u \in H^1([o,1] ; R^{2n}) \,|\, u(o) = u(1)\}$ and $Au = -J\dot{u}$ if $u \in dom(A)$. The continuous operator $F : H \to H$ is defined by $F(u)(t) := \nabla h(t,u(t)$ $u \in H$. Its potential $\Phi(u)$ is given by $\Phi(u) := \int_o^1 h(t,u(t))dt$, so that $F(u) = \nabla\Phi(u)$. Since $J^2 = -1$ we can write the equation (2) in the form $-J\dot{x} = \nabla h(t,x)$ and one sees that every solution $u \in dom(A)$ of the equation $Au = F(u)$ defines (by periodic continuation) a classical 1-periodic solution of (2). Conversely, every 1-periodic solution of (2) defines (by restriction) a solution u of the functional equation. With these notations the functional f defined by (3) becomes

$$f(u) = \frac{1}{2} < Au,u > - \Phi(u),$$

for $u \in dom(A)$. We look for critical points of f.

By assumption, there is a constant $\alpha > o$ such that

$$(4) \qquad |h''(t,x)| \leq \alpha$$

for all $(t,x) \in R \times R^{2n}$, where ' stands for the derivative in the x-variable. We shall use this estimate in order to reduce the problem of finding critical points of the functional f on dom(A) to the problem of finding critical points of a related functional, which is defined on a finite dimensional subspace of the Hilbert space H.

First observe that the operator A is selfadjoint, $A = A^*$. It has closed range and a compact resolvent. The spectrum of A, $\sigma(A)$, is a pure point spectrum and $\sigma(A) = 2\pi\,\mathbb{Z}$. Every eigenvalue $\lambda \in \sigma(A)$ has multiplicity 2n and the eigenspace $E(\lambda) := \ker(\lambda - A)$ is spanned by the orthogonal basis given by the loops:

$$t \to e^{t\lambda J} e_k = (\cos \lambda t) e_k + (\sin \lambda t) J e_k,$$

$k = 1,2,\dots,2n$, where $\{e_k \mid 1 \le k \le 2n\}$ is the standard basis in R^{2n}. In particular $\ker(A) = R^{2n}$; that is the kernel of A consist precisely of the constant loops in R^{2n}. Denoting by $\{E_\lambda \mid \lambda \in R\}$ the spectral resolution of A we define the orthogonal projection $P \in L(H)$ by

$$P = \int_{-\beta}^{\beta} dE_\lambda, \text{ with } \beta \ge 2\alpha, \quad (\alpha \text{ as in (4)})$$

where $\beta \notin 2\pi\,\mathbb{Z}$. Let $P^\perp = 1 - P$ and set $Z = P(H)$ and $Y = P^\perp(H)$. Then $H = Z + Y$ and $\dim Z < \infty$. With these notations the equation $Au - F(u) = o$, for $u \in \mathrm{dom}(A)$ is equivalent to the pair of equations

$$\text{(5)} \qquad \begin{aligned} APu - PF(u) &= o \\ AP^\perp u - P^\perp F(u) &= o. \end{aligned}$$

Now writing $u = Pu + P^\perp u = z + y \in Z + Y$ we shall solve, for fixed $z \in Z$, the second equation of (5) which becomes $Ay - P^\perp F(z+y) = o$. With $A_o := A|Y$ this equation is equivalent to

$$\text{(6)} \qquad Y = A_o^{-1} P^\perp F(z+y).$$

Observe that $|A_o^{-1}| \le \beta^{-1}$ and $|P^\perp| = 1$. Also, from (4) we conclude that $|F(u) - F(v)| \le \alpha|u-v|$ for all $u,v \in H$. Consequently, in view of $\beta \ge 2\alpha$, the right hand side of (6) is a contraction operator in H having contraction constant 1/2. We conclude, for fixed $z \in Z$, that the equation (6) has a unique solution $y = v(z) \in Y$. Since $(A_o^{-1} y)(t) = \int_o^t Jy(s)ds$, we have $A_o^{-1}(Y) \subset H^1$ and therefore $v(z) \in \mathrm{dom}(A)$. Moreover, the map $z \to v(z)$ from Z into Y is Lipschitz-continuous. In fact, we have $|v(z_1) - v(z_2)| \le \frac{1}{2}\{|z_1 - z_2| + |v(z_1) - v(z_2)|\}$. Setting

$$u(z) = z + v(z)$$

we now have to solve the first equation of (5), namely $Az - PF(u(z)) = o$.

One verifies readily that

$$\nabla g(z) = Az - PF(u(z)) \text{ with } g(z) := f(u(z)).$$

It remains to find critical points of the function g, which is defined on the finite dimensional space Z.

The critical points are now found by means of C. Conley's Morse-theory, which we recall next.

C. Conley's Index Theory [40]

To recall the classical Morse theory we consider a C^2-function $f: M \to R$, defined on a compact manifold M of dimension d. The critical points of f are the equilibrium points of the gradient-flow $\phi^t(x)$ of $\dot{x} = -\nabla f(x)$ on M. Assume that the critical points are isolated, so that $x_1, \ldots, x_m$ are all the critical points, $\nabla f(x_j) = o$. Then for every $x \in M$, there exists a pair of indices $i \leq j$ such that $\lim \phi^t(x) = x_i$ as $t \to +\infty$, $\lim \phi^t(x) = x_j$ as $t \to -\infty$, and $f(x_i) \leq f(x_j)$. This follows from the gradient structure of the flow. Therefore all the points $x \in M$ tend in forward and backward time to the critical points. If these are nondegenerate, then the rest points x_j are hyperbolic and the only topological local invariant of x_j is the dimension d_j of the unstable invariant manifold $W^-_{x_j} = W_j$ so that $\dim(W_j) = d_j$. Now $M = \bigcup_{1 \leq j \leq m} W_j$ and there is a relation between the global topological invariants of M, namely the Betti-numbers β_k, $o \leq k \leq d$, and the local invariants of the critical points. This relation is given by the following Morse-inequalities:

$$\sum_{j=1}^{m} t^{d_j} = \sum_{k=o}^{d} \beta_k \, t^k + (1+t)\, Q(t) \; ,$$

where Q is a polynomial having nonnegative integer coefficients. See for example R. Bott [26] and [27].

Conley's Index theory generalizes this Morse-theory to flows which are not necessarily gradient flows. Consider any topological flow $\phi^t(x) = x \cdot t$ on a locally compact metric space X. Then an index can be defined not only for isolated equilibrium points, but for every isolated invariant set $S \subset X$. The invariant set S is called isolated, if it is the maximal invariant set, $I(N)$, of a compact neighborhood N of itself, $S = I(N) \subset \text{int}(N)$. The index will be the homotopy type $h(S)$ of a pointed compact topological space. It is defined by means of an index pair (N_1,N_o) of S, where the compact pair $N_o \subset N_1$ measures some of the qualitative behaviour of the flow in a neighborhood of S, N_o is the exit set. The precise definition is as follows:

Definition (Index pair)

Let S be a compact, isolated invariant set of a continuous flow. A compact pair (N_1,N_o) is an index pair for S if:

(i) $cl(N_1 \setminus N_o)$ is an isolating neighborhood for S.
(ii) N_o is positively invariant relative to N_1, i.e. if $x \in N_o$ and $x\cdot[o,t] \subset N_1$ then $x\cdot[o,t] \subset N_o$.
(iii) If $x \in N_1$ and $x\cdot\mathbb{R}^+ \not\subset N_1$ then there is a $t \geq o$ such that $x\cdot[o,t] \subset N_1$ and $x\cdot t \in N_o$.

Theorem 11

There exists an index pair for a compact, isolated invariant set S. *If* (N_1,N_o) *and* $(\overline{N}_1,\overline{N}_o)$ *are two index pairs for* S, *then* $({}^{N}1/N_o,*) \cong \cong ({}^{\overline{N}}1/\overline{N}_o,*)$ *are homotopically equivalent. The homotopy class*

$$h(S) = [({}^{N}1/N_o,*)]$$

is called the index of S, *here* (N_1,N_o) *is any index pair for* S.

Crucial for the applications is the follwoing continuation-theorem:

Theorem 12

Let ϕ_λ, $o \leq \lambda \leq 1$, *be a continuous* 1-*parameter family of flows. Let* N *be a compact neighborhood which is isolating for every* λ, *i.e.* $I(N|\phi_\lambda) =: S_\lambda \subset \text{int}\,(N)$. *Then* $h(S_\lambda|\phi_\lambda) = h(S_o|\phi_o)$ for $o \leq \lambda \leq 1$.

The setting for the generalized Morse-theory is described by the following definition, which reminds of the gradient flow on a manifold $M = S$ and the critical points $\{x_j\} = M_j$.

Definition:

A Morse-decomposition of S is an ordered family $\{M_1,\dots,M_m\}$ of disjoint, compact and invariant subsets of S, such that the following property holds. If $x \in S \setminus \bigcup_{j=1}^{m} M_j$ then there is a pair of indices $i < j$

such that the positive limit set of x as $t \to +\infty$, and the negative limit set as $t \to -\infty$ satisfy:

$$\omega(x) \subset M_i \quad \text{and} \quad \omega^*(x) \subset M_j.$$

Define the algebraic invariants of an isolated invariant set S as $p(t,h(S)) := \sum_{j \geq 0} \gamma_j t^j$, where $\gamma_j = \text{rank } H^j(N_1,N_0)$, with (N_1,N_0) being any index pair for S. The relation between the invariants of S and the local invariants of a Morse-decomposition of S is given by the following theorem, proved in [41].

Theorem 13

If $\{M_1,\ldots,M_m\}$ *is an ordered Morse-decomposition of an isolated, compact invariant set* S, *then*

$$\sum_{j=1}^{m} p(t,h(M_j)) = p(t,h(S)) + (1+t)\, Q(t),$$

where Q *is a formal power series having nonnegative integer coefficients.*

We point out, that this theory has been extended to semiflows on metric spaces which are not necessarily locally compact, see [91]. •

The application of this outlined theory to the gradient flow ∇g on Z is as follows. We let S be the set of all bounded solutions of the gradient flow. Since the Hamiltonian system is asymptotically linear and since the linear system at infinity is nondegenerate one concludes, that S is compact, hence isolated and has, therefore an index. In order to compute this index, the invariance of the index under deformation is crucial. By definition of the index j_∞ of the linear system at ∞ there is a deformation to a corresponding system with constant coefficients, in such a way that the set S remains isolated in a big ball. Using the continuation theorem it is then easily proved, that the index of S is that of a hyperbolic rest point in Z, which is a pointed sphere having the dimension of the unstable invariant manifold. Also, the Maslov-indices of the periodic solutions are related to the Conley Indices of the corresponding critical points of g, and are easily computed in case the periodic solutions are nondegenerate. We summarize these observations in a

Lemma

Assume that the Hamiltonian function mets the assumptions of theorem 9. Then

(i) The set S *of bounded solutions of* $\dot{z} = \nabla g(z)$ *is compact, hence has an index. It is the homotopy type of a pointed sphere:* $h(S) = [\dot{S}^{m_\infty}]$ *with* $m_\infty = \frac{1}{2}\dim Z - j_\infty$. *Therefore* $p(t,h(S)) = t^{m_\infty}$.

(ii) If $x(t)$ *is a nondegenerate T-periodic solution with index* j, *then the corresponding critical point* z *of* g *is an isolated invariant set, and* $h(\{z\}) = [\dot{S}^m]$, *where* $m = \frac{1}{2}\dim Z - j$; *hence* $p(t,h(\{z\})) = t^m$.

By the Lemma the invariant set $S \subset Z$ is compact and of homotopy type $h(S) = [\dot{S}^{m_\infty}]$ with $m_\infty = d - j_\infty$. This is not the index of the empty set which is a pointed one point space hence has the homotopy type $[(\{p\},p)]$ for an arbitrary point p. Therefore $S \neq \emptyset$ and because the limit set of a bounded orbit of a gradients system consists of critical points, the function a possesses at least one critical point and consequently the Hamiltonian equation admits at least one T-periodic solution. If the periodic orbit found above is nondegenerate, it has an index denoted by $j \in Z$. The corresponding critical point z of g is then, by the Lemma, an isolated invariant set with index $h(\{z\}) = [\dot{S}^m]$, where $m = d - j$. Assume z is the only critical point of g, then $S = \{z\}$, since we are dealing with a gradient system and therefore $h(S) = [\dot{S}^m]$ which, on the other hand is equal to $[\dot{S}^{m_\infty}]$ and consequently $m = m_\infty$. Therefore if $j \neq j_\infty$ and hence $m \neq m_\infty$ there must be more than one critical point of a. Assume now that the Hamiltonian system possesses two nondegenerate periodic orbits having indices j_1 and j_2. We claim that there is at least a third periodic orbit. In fact, if this is not the case, then the isolated invariant set S contains precisely two isolated critical points z_1 and z_2 with indices $h(\{z_1\}) = [\dot{S}^{m_1}]$, $m_1 = d - j_1$ and $h(\{z_2\}) = [\dot{S}^{m_2}]$, $m_2 = d - j_2$. If we label them such that $g(z_1) \leq g(z_2)$, then (z_1, z_2) is an admissible Morse-decomposition of S. From theorem 13 we conclude the identity $p(t,h(\{z_1\})) + p(t,h(\{z_2\})) = p(t,h(S)) + (1+t)\,Q(t)$, which, by the Lemma, leads to the identity $t^{m_1} + t^{m_2} = t^{m_\infty} + (1+t)\,Q(t)$. Setting $t = 1$ we find the equation $2 = 1 + 2Q(1)$ with a nonnegative integer $Q(1)$. This is nonsense, hence we must have at least three critical points of g.

Assume finally all the periodic solutions to be nondegenerate and denote their indices by j_k, $k=1,2,\ldots$ They correspond to the critical points of g, which are isolated. Since C is compact there are only finitely many of them, say $(z_1,\dots,z_n)$. We order them such that $g(z_i) \le g(z_j)$ if $i < j$. Then $(z_1,\dots,z_n)$ is an admissible ordering of a Morse decomposition of S, and by Theorem 13 and the Lemma we have $\sum_{k=1}^{n} p(t,h(z_k)) = t^{m_\infty} + (1+t)\, Q(t)$, with $m_\infty = d - j_\infty$. By assumption the periodic solutions are nondegenerate, hence by the Lemma $p(t,h(z_k)) = t^{m_k}$, $m_k = d - j_k$, so that

$$\sum_{k=1}^{n} t^{m_k} = t^{m_\infty} + (1+t)\, Q(t),$$

which after multiplication by t^{-d}, $d = \frac{1}{2} \dim Z$ becomes the advertized identity in Theorem 9. We conclude that there is at least one periodic solution having index j_∞. Also, setting $t = 1$ we find $n = 1 + 2\, Q(t)$, hence the number of periodic solutions is odd as claimed in Theorem 9. This finishes the proof of Theorem 9. •

We now turn to a very special result.

2. Hamiltonian vectorfields on a torus

Finally, we consider a Hamiltonian system on the torus $T^{2n} = \mathbb{R}^{2n}/Z^{2n}$, which, on the covering space R^{2n}, is given by

$$(7) \qquad \dot{x} = J\nabla h(t,x), \qquad x \in R^{2n},$$

with h periodic in all the variables of period 1. Then the following statement is proved in [43].

Theorem 14

The Hamiltonian vectorfield (7) *on* T^{2n}, *with the function* $h(t,x) \in C^2(R \times R^{2n})$ *being periodic of period* 1 *possesses at least* 2n+1 *periodic solutions of period* 1.

The periodic solutions found by the theorem are contractible loops on T^{2n}, i.e. are given as periodic functions on R^{2n}. One expects more periodic solutions, if all the periodic solutions are known to be nondegenerate. Recall that a 1-periodic solution is nondegenerate, if it has no Floquet-multiplier equal to 1. Recall that $\lambda \in C$ is a Floquet-multiplier of a periodic solution $x(t) = x(t+1)$, if λ is an eigenvalue of $d\phi^1(x(o))$, where ϕ^t is the flow of the corresponding timeindependent vectorfield. Indeed, the following statement holds true:

Theorem 15

Assume that all the periodic solutions having period 1 of the system (7) are nondegenerate, then there are at least 2^{2n} of them.

From these existence statements for periodic solutions one deduces immediately the following Corollary for the symplectic map $\psi = \phi^1$. We call a fixed point $x = \psi(x)$ nondegenerate if 1 is not an eigenvalue of $d\psi(x)$.

Corollary 1.

Every symplectic C^1-diffeomorphism ψ on the torus $T^{2n} = R^{2n}/Z^{2n}$, which is generated by a globally Hamiltonian vectorfield, possesses at least $2n+1$ fixed points: If, moreover, all the fixed points of ψ are nondegenerate, then there are at least 2^{2n} of them.

The symplectic diffeomorphism which meet the assumption of the Corollary can be characterized as follows. If (M,ω) is any compact, symplectic and smooth manifold, we denote $\mathrm{Diff}^{\infty}(M,\omega)$ the topological group of symplectic C^{∞}-diffeomorphisms ψ, i.e. $\psi^*\omega = \omega$. Let $\mathrm{Diff}^{\infty}(M,\omega)$ be the identity component in $\mathrm{Diff}^{\infty}(M,\omega)$, which can be shown to be the identity component by smooth arcs in $\mathrm{Diff}^{\infty}(M,\omega)$. It has been proved by A. Banyaga [16], that the commutator-subgroup of $\mathrm{Diff}_o^{\infty}(M,\omega)$ consists precisely of those symplectic diffeomorphisms, which are generated by globally Hamiltonian vectorfields on M, and hence agrees with the subgroup of symplectic diffeomorphisms having vanishing socalled Calabi-invariant.

As a special case we consider a measure preserving diffeomorphism of T^2, which is homologeous to the identity map on T^2 and hence is, on the covering space R^2, of the form

$$(8) \qquad \psi : x \to x + f(x), \quad x \in R^2$$

with f being periodic. As observed by V.I. Arnold, see [9], this map ψ is generated by a globally Hamiltonian vectorfield on T^2 if and only if the meanvalue of f over the torus vanishes, i.e. $[f] = o$. A proof can be found in [43]. We therefore conclude from theorem 1 the following result, which was conjectured by V.I. Arnold in [8] and [9].

Corollary 2.

Every measure preserving C^1-diffeomorphism of T^2 which is of the form (8) with $[f] = o$ has at least 3 fixed points.

The condition $[f] = o$ is clearly necessary in order to guarantee a fixed point, as the translation map $x \to x + c$ shows, which has no fixed points on T^2, if $c \notin Z^2$. It should be emphasized that the symplectic map is not assumed to be C^1-close to the identity map. Under this additional assumption the above fixed points are immediately found as critical points of a so called generating function, see [9]. In fact, write the map (8) explicitely in the form

$$X = x + p(x,y)$$
$$Y = y + q(x,y)$$

with p,q periodic. Since, by assumption, $dX \wedge dY = dx \wedge dy$ we conclude that the following one form

$$(X-x)\ (dY+dy) - (Y-y)\ (dX+dx) = dS(x,y)$$

is closed, hence, on $\mathbb{R}^{2n}$, exact. The one form dS is periodic, hence $S(x,y)=c_1x+c_2y+s(x,y)$, s being periodic. One verifies readily that $c_1= =c_2=o$ if and only if $[p]=[q]=o$, in which case S in a function on T^2. It has at least 3 critical points, which are obviously fixed points of the map ψ provided the two 1-forms $(dY+dy)$ and $(dX+dx)$ are linearly independent at the critical points of S. This, on the other hand, is the case if and only if (-1) is not an eigenvalue of $d\psi$ at the critical points of S, hence in particular if ψ is C^1-close to the identity map. The idea of relating fixed points of symplectic maps to critical points of a related function on the corresponding manifold goes back to H. Poincaré [64]. It has been exploited by A. Banyaga [17], J. Moser [63] and A. Weinstein [77] in order to guarantee fixed points for symplectic maps, which are however always assumed to be C^1-close to the identity map.

The idea of the proof of theorem 14 is as follows. The periodic orbits correspond to the critical points $\nabla g(z) = o$, where, however, this time ∇g is a gradient flow on the manifold $M = T^{2n} \times \mathbb{R}^N \times \mathbb{R}^N$ for some large N, this follows from the periodicity of h. From the fact, that h and its derivatives are uniformly bounded it follows, that the set of bounded solutions of this gradient flow is compact and contained in the compact set $B := T^{2n} \times D \times D$, where D is a disc in R^N. Moreover $B^- := T^{2n} \times \partial D \times D$ is the exit set and $B^+ = T^{2n} \times D \times \partial D$ is the entrance set, so that B is an isolating block in the sense of [40]. The proof now follows from two general statements for general flows, which are not necessarily gradient flows. First consider any continuous flow which admits the above very special isolating block B, with exit set B^- and entrance set B^+. Then the invariant S of the flow contained in B carries cohomology which it obtains from the torus T^{2n}. In fact it can be shown:

$$\ell(S) \geq \ell(B) = \ell(T^{2n}) = 2n+1 ,$$

where $\ell(X)$ denotes the cup long of a compact space X. The second statement concerns Morse-decompositions. If $\{M_1,\dots,M_k\}$ is an ordered Morse-decomposition of a compact, isolated invariant set S of a continuous flow, then

$$\ell(S) \leq \sum_{j=1}^{k} \ell(M_j).$$

If, in addition, the flow on S is gradientlike with finitely many rest points then these rest points are a Morse-decomposition of S. In this case $\ell(M_j) = 1$ and we obtain the estimate

$$\ell(S) \leq \sum_{j=1}^{k} 1 = \#\{\text{rest points}\}.$$

We conclude that the gradient flow ∇g possesses at least $2n+1$ rest points and theorem 14 follows. •

Theorem 15 follows immediately from the Morse-equation (Theorem 13). Namely for the set S of bounded solutions of ∇g on M, one shows that $p(t,h(S)) = \sum_{j=o} \binom{2n}{j} t^{N+j}$, since $\dim H^j(T^{2n}) = \binom{2n}{j}$, so that the Morse inequalities become, in case of nondegenerate critical points $z_1,\dots,z_m$:

$$\sum_{j=1}^{m} t^{d_j} - \sum_{k=o}^{2n} \binom{2n}{k} t^{N+k} = (1+t)\, Q(t),$$

and therefore $m \geq \sum_{k=o}^{2n} \binom{2n}{k} = 2^{2n}$, as claimed. •

References:

[1] J.C. Alexander / J.A. Yorke: "Global Bifurcation of periodic orbits", Amer. J. Math. 100 (1978), 263-292

[2] H. Amann / E. Zehnder: "Nontrivial Solutions for a Class of Nonresonance Problems and Applications to Nonlinear Differential Equations", Annali Sc. Norm. Sup. Pisa, Serie IV, Vol. VII (1980), 539-603

[3] H. Amann / E. Zehnder: "Periodic solutions of Asymptotically linear Hamiltonian systems", Manus. Math. 32, (1980), 149-189

[4] A. Ambrosetti: "Recent Advances in the study of the existence of periodic orbits of Hamiltonian systems", Preprint, SISSA, Trieste (1982), 1-19

[5] A. Ambrosetti / G. Mancini: "Solutions of Minimal period for a Class of Convex Hamiltonian systems", Math. Ann. 255 (1981), 405-421

[6] A. Ambrosetti / G. Mancini: "On a theorem by Ekeland und Lasry concerning the number of periodic Hamiltonian trajectories", J. Diff. Equ. 43 (1982), 249-256

[7] A. Ambrosetti / P.H. Rabinowitz: "Dual variational methods in critical point theory and applications", Journal Functional Analysis 14, (1973), 349-381

[8] V.I. Arnold: Proceedings of Symposia in Pure Mathematics, Vol. XXVIII A.M.S. (1976), p.66

[9] V.I. Arnold: "Mathematical Methods of Classical Mechanics", (Appendix 9), Springer 1978

[10] A. Bahri / H. Berestycki: "Points critiques de perturbations de fonctionnelles paires et applications", C.R. Acad. Sc. Paris, t. 291 série A (1980), 189-192

[11] A. Bahri / H. Berestycki: "Existence d'une infinité de solutions périodiques de certains systèmes hamiltoniens en présence d'un

terme de contrainte",C.R. Acad. Sc. Paris, t. 292, série A(1981), 315-318

[12] W. Ballmann / G. Thorbergsson / W. Ziller: "Closed geodesics on positively curved manifolds", Annals of Math. 116, (1982), 231-247

[13] W. Ballmann / W. Ziller: "On the number of closed geodesics on a compact Riemannian manifold", Duke Math. J. 49 (1982), 629-632

[14] V. Bangert: "Closed Geodesics on Complete Surfaces", Math. Ann. 251 (1980), 83-96

[15] V. Bangert / W. Klingenberg: "Homology generated by iterated closed geodesics", Preprint Freiburg, Bonn, 1981

[16] A. Banyaga: "Sur la structure du groupe des difféomorphismes qui presêrvent une forme symplectique", Comment. Math. Helvetici 53 (1978), 174-227

[17] A. Banyaga: "On fixed points of symplectic maps", Preprint

[18] V. Benci:"Some critical point theorems and applications",Comm. Pure Appl. Math. 33 (1980)

[19] V. Benci / P.H. Rabinowitz: "Critical point theorems for indefinite functionals", Inv. math. 52 (1979), 336-352

[20] M. Berger: "On periodic solutions of second order Hamiltonian systems", J. Math. Anal. Appl. 29 (1970), 512-522

[21] M. Berger: "Periodic solutions of second order dynamical systems and isoperimetric variational problems", Amer. J. Math. 93 (1971) 1-10

[22] M. Berger: "On a family of periodic solutions of Hamiltonian systems", J. Diff. Eq. 10 (1971), 17-26

[23] G.D. Birkhoff:"An extension of Poincaré's last geometric theorem", Acta Math. 47 (1925), 297-311

[24] G.D. Birkhoff: "Une generalization à n-dimensions du dernier théorème de géometrie de Poincaré", Comp. Rend. Acad. Sci. 192, (1931), 196-198

[25] G.D. Birkhoff / D.C. Lewis: "On the periodic motions near a given periodic motion of a dynamical system", Ann. Mat. Pura Appl. 12 (1933), 117-133

[26] R. Bott: "Marston Morse and his mathematical works", Bull (New Series) A.M.S. 3 (1980), 907-950

[27] R. Bott: "Lectures on Morse theory, old and new", Bulletin (New Series) A.M.S. 7 (2) (1982), 331-358

[28] M. Bottkol: "Bifurcation of periodic orbits on manifolds and Hamiltonian systems, Thesis, New York University 1977

[29] H. Brezis / J.M. Coron / L. Nirenberg: "Free vibrations for a nonlinear wave equation and a theorem of P. Rabinowitz", Comm. Pure Appl. Math 33 (1980), 667-689

[30] M. Brown / W.D. Neumann: "Proof of the Poincaré-Birkhoff fixed point theorem", Michigan Math. Journ. 24 (1977), 21-31

[31] K.C. Chang: "Solutions of asymptotically linear operator equations via Morse theory", Comm. Pure and Appl. Math 34 (1981), 693-712

[32] A. Chenciner: "Points périodiques de longues periodes au voisinage d'une bifurcations de Hopf dêgenerée de difféomorphismes de $\mathbb{R}^2$", Preprint 1982

[33] S.N. Chow / J. Mallet-Paret: "The Fuller Index and Global Hopf Bifurcation", J. of Diff. Equ. 29 (1978), 66-85

[34] S.N. Chow / J. Mallet-Paret: "Periodic solutions near an equilibrium of a nonpositive definite Hamiltonian system", Michigan State Univ. Preprint

[35] S.N. Chow / J. Mallet-Paret / J.A. Yorke: "Global Hopf Bifurcation from a multiple eigenvalue", Nonlinear Analysis, Theory, Meth. and Appl. 2 (1978), 753-763

[36] F. Clarke: " A classical variational principle for periodic Hamiltonian trajectories", Proc. Amer. Math. Soc. 76 (1979), 186-188

[37] F. Clarke: "Periodic solutions to Hamiltonian inclusions", J. Diff. Equ. 40 (1981), 1-6

[38] F. Clarke / I. Ekeland: "Nonlinear Oscillations and Boundary-Value Problems for Hamiltonian systems", Archive Rat. Mech. and Analysis, in press

[39] F. Clarke / I. Ekeland: "Hamiltonian Trajectories Having Prescribed Minimal Periods", Comm. on Pure and Appl. Math. 33 (1980), 103-116

[40] C.C. Conley: "Isolated invariant sets and the Morse index", CBMS Regional Conf. Series in Math 38 (1978) A.M.S. Providence R.I.

[41] C.C. Conley / E. Zehnder: "Morse type index theory for flows and periodic solutions for Hamiltonian equations", to appear in Comm. Pure and Appl. Math.

[42] C.C. Conley / E. Zehnder: " An index theory for periodic solutions of a Hamiltonian system", to appear in the Proceedings of the Rio Conference on Dynamical systems

[43] C.C. Conley / E. Zehnder: "The Birkhoff-Lewis fixed point theorem and a conjecture of V. Arnold", Preprint FIM, ETH Zürich (1982), 1-26

[44] H. Duistermaat: "On periodic solutions near equilibrium points of conservative systems", Arch. Rat. Mech. Anal. 45 (1972), 143-160

[45] H. Duistermaat: "On the Morse Index in variational calculus", Adv. in Math. 21 (1976), 173-195

[46] H. Duistermaat: "Periodic solutions near equilibrium points of Hamiltonian systems", Utrecht, Dept. of Math., Preprint Nr. 156 (1980)

[47] I. Ekeland: "Periodic solutions of Hamiltonian equations and a

theorem of P. Rabinowitz", J. Differential Equations 34 (1979), 523-534

[48] I. Ekeland: "La théorie des perturbations au voisinage des systèmes Hamiltonian convexes", École Polytechnique, Centre de Mathématiques, Exposé n° VII (1981)

[49] I. Ekeland: "Oscillations de systèmes Hamiltoniens non linéaires III", Bull. Soc. Math. France 109 (1981), 297-330

[50] I. Ekeland: "Forced oscillations for Nonlinear Hamiltonian Systems II" in Advances in Mathematics, volume en l'honneur de Laurent Schwartz, Nachbin, éd. 1981, Academic Press

[51] I. Ekeland: "Dualité et stabilité des systèmes hamiltoniens", C.R. Acad. Sc. Paris 294 Série I (1982) 673-676.

[52] I. Ekeland / J.M. Lasry: "On the number of periodic trajectories for a Hamiltonian flow on a convex energy surface", Ann. of Math. 112 (1980), 283-319

[53] E.R. Fadell / P.H. Rabinowitz: "Generalized cohomological index theories for Lie group actions with an application to bifurcation questions for Hamiltonian systems", Inv. Math. 45 (1978), 139-174

[54] H. Gluck / W. Ziller: "Existence of periodic motions of conservative systems", University of Pennsylvania (1980), Preliminary draft

[55] W.B. Gordon: "A theorem on the existence of periodic solutions to Hamiltonian systems with convex potentials", J. Diff. Eq. 10 (1971), 324-335

[56] T.C. Harris: "Periodic solutions of arbitrary long period in Hamiltonian systems", J. Diff. Eq. 4 (1968), 131-141

[57] J. Horn: "Beiträge zur Theorie der kleinen Schwingungen", Zeit. Math. Phys. 48 (1903), 400-434

[58] H. Jacobowitz: "Periodic solutions of $x'' + f(x,t) = o$ via the Poincaré-Birkhoff theorem", J. Diff. Eq. 20 (1976), 37-52

[59] W. Klingenberg: "Lectures on closed geodesics", Grundlehren Vol. 230, Springer 1978

[60] A. Lyapunov: "Problème générale de la stabilite du mouvement", Ann. Fac. Sci. Toulouse (2) (1907), 203-474

[61] J. Moser: "Periodic orbits near an equilibrium and a theorem by Alan Weinstein", Comm. Pure Appl. Math. 29 (1976), 727-747

[62] J. Moser: "Proof of a generalized form of a fixed point theorem due to G.D. Birkhoff", Springer Lecture Notes in Mathematics, Vol. 597: Geometry and Topology (1977), 464-494

[63] J. Moser: "A fixed point theorem in symplectic geometry", Acta Math. 141 (1978), 17-34

[64] H. Poincaré: "Méthodes nouvelles de la mécanique célèste", Vol. 3, chap. 28, Gauthier Villars, Paris (1899)

[65] J. Pöschel:"Integrability of Hamiltonian Systems on Cantor Sets", Comm. Pure Appl. Math. 35 (1982) 653-696

[66] P.H. Rabinowitz: "A variational method for finding periodic solutions of differential equations", Nonlinear Evolution Equations (M.G. Crandall, editor), Academic Press (1978), 225-251

[67] P.H. Rabinowitz: "Some minimax theorems and applications to nonlinear partial differential equations", Nonlinear Analysis, A Collection of Papers in Honor of Erich H. Rothe, 161-177, Academic Press 1978

[68] P.H. Rabinowitz: "Periodic solutions of Hamiltonian systems", Comm. Pure Appl. Math. 31 (1978), 157-184

[69] P.H. Rabinowitz: "Periodic solutions of a Hamiltonian system on a prescribed energy surface", J. Diff. Eq. 33 (1979), 336-352

[70] P.H. Rabinowitz: "Periodic solutions of Hamiltonian systems: a survey", SIAM J. Math. Anal. 13 (1982), 343-352

[71] P.H. Rabinowitz: "On periodic solutions of large norm of some ordinary and partial differential equations", Ergodic Theory and Dynamical Systems, Proc. Sp. Yr.-Maryland 79-80, A. Katok, ed., to appear

[72] P.H. Rabinowitz: "Subharmonic solutions of Hamiltonian systems", Comm. Pure Appl. Math. XXXIII (1980), 609-633

[73] J.A. Sanders: "Are higher order resonances really interesting?", Celestial Mech. 16 (1978), 421-440

[74] D.S. Schmidt: "Periodic solutions near a resonant equilibrium of a Hamiltonian system", Celestial Mech. 9 (1974), 81-103

[75] H. Seifert: "Periodische Bewegungen mechanischer Systeme", Math. Z. 51 (1948), 197-216

[76] A. Weinstein: "Normal modes for nonlinear Hamiltonian systems, Inv. Math 20 (1973), 47-57

[77] A. Weinstein: "Lectures on symplectic manifolds", CBMS, Regional conf. series in Math. 29 (1977)

[78] A. Weinstein: "Bifurcations and Hamilton's Principle", Math. Z. 159 (1978), 235-248

[79] A. Weinstein: "Periodic orbits for convex Hamiltonian systems", Ann. Math. 108 (1978), 507-518

[80] A. Weinstein: "On the hypotheses of Rabinowitz' periodic orbit theorems", J. Diff. Eq. 33 (1979), 353-358

[81] M. Willem: "On the number of Periodic Hamiltonian orbits on a convex surface", Preprint (1983)

[82] H. Amann: "Gewöhnliche Differentialgleichungen" Teubner 1983

[83] P.A. Schweitzer: "Counterexamples to the Seifert conjecture and opening closed leaves of foliations", Annals of Math. 100 (1974), 386-400

[84] T.W. Wilson: "On the minimal sets of nonsingular vectorfields", Annals of Math. 84 (1966), 529-536

[85] D. Clark: "On periodic solutions of autonomous Hamiltonian systems of ordinary differential equations", Proc. AMS 39 (1973), 579-584

[86] A. Bahri / H. Berestycki: "Forced vibrations of super-quadratic Hamiltonian systems", Preprint, Univ. Pierre et Marie Curie (1982)

[87] I. Ekeland: "A perturbation theory near convex Hamiltonian systems", Technical Report 82-1 (1982)

[88] C.B. Croke / A. Weinstein: "Closed Curves on Convex Hypersurfaces and Periods of Nonlinear Oscillations", Preprint

[89] A. Weinstein: "Symplectic V-Manifolds, Periodic orbits of Hamiltonian Systems, and the Volume of certain Riemannian Manifolds", Comm. on Pure and Appl. Math. 30 (1977), 265-271

[90] J. Moser: "Addendum to 'Periodic orbits near an Equilibrium and a Theorem by A. Weinstein' ", Comm. Pure and Appl. Math. 31 (1978), 529-530

[91] K.P. Rybakowski / E. Zehnder: "A Morse-Equation in Conley's Index theory for semiflows on metric spaces", to be published in Dynamical Systems and Ergodic Theory

[92] P.H. Carter: "An Improvement of the Poincaré-Birkhoff Fixed point theorem", Transactions AMS 269 (1982), 285-299

[93] A. Chenciner: "Sur un énoncé dissipatif du théorème géométrique de Poincaré-Birkhoff", C.R. Acad. Sc. Paris 294 I (1982), 243-245

[94] E. Zehnder: "Homoclinic Points near elliptic Fixed Points", Comm. Pure and Appl. Math. XXVI (1973) 131-182

[95] E. Zehnder: "Generalized Implicit Function Theorems with Applications to Some Small Divisor Problems I, and II", Comm. Pure and Appl. Math. XXVIII (1975) 91-140 and XXIX (1976) 49-111

[96] A. Hofer: "A new proof for a result of Ekeland and Lasry concerning the Number of periodic Hamiltonian trajectories on a prescribed energy surface", Boll. U.M.I. (6), 1-B (1982) 931-942

[97] C.L. Siegel / J. Moser: "Lectures on Celestial Mechanics", Springer Grundlehren Bd. 187 (1971)

[98] J. Moser: "Stable and Random Motions in Dynamical Systems", Annals of Math. Studies, Vol. 77, Princeton University Press (1973)

[99] C.C. Pugh / R.C. Robinson: "The C^1 Closing Lemma, including Hamiltonians", Preprint

[100] J.V. Ralston: "On the construction of quasimodes associated with stable periodic orbits", Comm. Math. Phys. 51 (1976) 219-242

[101] Y. Colin de Verdière: Compositio Math. 27 (1973) 83-106 and 159-184

[102] J. Mather: "Existence of Quasiperiodic Orbits for Twist Homeomorphisms of the Annulus", to appear in Topology

[103] J. Mather: "A Criterion for the Nonexistence of invariant Circles", Preliminary draft (1982)

[104] M. Herman: "Contre examples de classe $C^{3-\varepsilon}$ et à nombre de rotation fixé au théorème des courbes invariantes", (1979), to be published

[105] A. Katok: "Some Remarks on Birkhoff and Mather twist map theorems" (1982) to be published in Dynamical systems and ergodic theory

[106] S. Albeverio / P. Blanchard / R. Höegh-Krohn: "Feynman path integrals and the trace formula for Schrödinger operators", Commun. math. Phys. 83 (1982) 49-76

[107] P. Hartman: "On boundary value problems for superlinear second

order differential equations", Jour. Diff. Eq. 26 (1977) 37-53

[108] V. Benci: "A geometrical index for the group S^1 and some applications to the research of periodic solutions of ordinary differential equations", to appear

[109] V. Benci: "On the critical point theory for indefinite functionals in the presence of symmetries", to be published.

This paper has been written while the author was a member at the Forschungsinstitut ETH Zürich. The visit was made possible by a Akademie-Stipendium of the Stiftung Volkswagenwerk.